Synthesis Lectures on Engineering, Science, and Technology

The focus of this series is general topics, and applications about, and for, engineers and scientists on a wide array of applications, methods and advances. Most titles cover subjects such as professional development, education, and study skills, as well as basic introductory undergraduate material and other topics appropriate for a broader and less technical audience.

Erik Cuevas · Daniel Zaldivar · Ernesto Ayala ·
Óscar González · Fernando Vega

DC Motors

Modeling, Designing and Building with 3D Printers

Erik Cuevas
University of Guadalajara
Guadalajara, Mexico

Daniel Zaldivar
University of Guadalajara
Guadalajara, Mexico

Ernesto Ayala
University of Guadalajara
Guadalajara, Mexico

Óscar González
University of Guadalajara
Guadalajara, Mexico

Fernando Vega
University of Guadalajara
Guadalajara, Mexico

ISSN 2690-0300   ISSN 2690-0327 (electronic)
Synthesis Lectures on Engineering, Science, and Technology
ISBN 978-3-031-64353-8   ISBN 978-3-031-64354-5 (eBook)
https://doi.org/10.1007/978-3-031-64354-5

This Springer imprint is published by the registered company Springer Nature Switzerland AG
The registered company address is: Gewerbestrasse 11, 6330 Cham, Switzerland

If disposing of this product, please recycle the paper.

Preface

DC motors are crucial components in various applications due to their simplicity, efficiency, and reliability. They convert direct electrical energy into mechanical energy, making them ideal for tasks that require precise speed control and high torque at low speeds. DC motors are extensively used in electric vehicles, robotics, and industrial machinery. In electric vehicles, they provide the necessary torque for acceleration and smooth speed control, enhancing performance and efficiency. In robotics, their precise control allows for accurate movement and positioning, which is essential for automation and complex tasks. Additionally, in industrial applications, DC motors drive conveyor belts, pumps, and other equipment, ensuring consistent and reliable operation. Their versatility and ability to deliver controlled performance make DC motors indispensable in modern technology and industry.

Designing a motor for a specific application is often complicated due to the advanced mechanical design skills needed to work with the metal components traditionally used in their construction. To simplify this process, plastic materials and 3D printers have emerged as viable alternatives for motor manufacturing. Utilizing these methods makes the design process easier and more cost-effective. Plastics can be molded with high precision, and 3D printing allows for rapid prototyping and customization. This approach reduces the expertise required in metalworking, enabling more individuals to engage in motor design and innovation. Consequently, this shift towards plastic materials and 3D printing not only lowers production costs but also democratizes the design process, fostering greater creativity and experimentation in motor development.

The goal of this book is to provide readers with the knowledge and practical skills necessary to understand, design, and construct their own functional DC motors using 3D printing technology. The book will provide a clear and accessible introduction to the fundamental concepts of DC motors. It will explain how they work, their different types, and their applications in a way that is easy for readers with limited technical background to understand. The book bridges the gap between theoretical knowledge and practical application. Readers can see how theoretical concepts translate into real-world devices. The book will guide readers through the process of building their own DC motors using 3D printing technology. By including step-by-step instructions, illustrations, and diagrams, readers will have a hands-on experience in creating functional devices. The focus on 3D printing allows readers to explore the field of customization. They can adapt the motor design to fit their specific needs, whether it's for a project, a prototype, or an application.

The material has been compiled from a teaching perspective. For this reason, the book is primarily intended as a textbook for undergraduate and postgraduate students of Science, Electrical Engineering. The book can be appropriated for Individuals who enjoy hands-on projects and have a general interest in electronics and mechanics. They may be looking to learn more about DC motors and how to build and customize them for personal projects.

The book is composed of seven chapters. Chapter 1 considers how permanent magnets, the winding, and its core, through their physical or electromagnetic properties, can generate magnetic fields that produce motion. As a result, concepts such as magnetic field, magnetic flux density, magnetic forces, electromagnetic induction, and Ampere's and Faraday's law will be presented. Additionally, we will explain the electromagnetic principles that enable the conversion of electrical energy into motion through an electric motor, as well as its component parts. We will focus on the analysis of magnetic circuits, which are metallic structures that can be wound with a conductor to produce a magnetic flux or utilize permanent magnets, which can maintain a state of magnetization without external energy. Furthermore, we will discuss the calculation of magnetic flux density, fundamental elements of an electromagnet such as the core, coils, and air spaces, as well as hysteresis and magnetic saturation properties.

Chapter 2 delves into the complex interplay between magnetic field dynamics, electrical circuit principles, and mechanical motion. It elucidates the essential process of energy conversion present in various systems. The investigation commences by analyzing how magnetic fields interact with electric currents to create forces and motions, a concept fundamental to technologies such as electric motors and generators. The chapter will elucidate the laws of electromagnetism, including those formulated by Faraday and Lenz, which explicate how modifications in magnetic fields engender electrical currents in circuits, and how these induced currents can yield mechanical forces.

Chapter 3 focuses on explaining the complex concepts required to identify optimal combinations of pole and slot counts for three-phase motors. Additionally, it explores ways to determine the appropriate winding layout once a suitable combination has been found. The chapter's structure provides clear instructions on optimizing motor design for superior performance and efficiency, particularly for advanced applications of brushless permanent magnet motors.

Chapter 4 focuses on providing an overview of the tools available in the widely used CAD software, Fusion 360. The chapter is designed to offer relevant data, examples, and advice on the limitations of 2D sketch operation when applied to the design of objects for 3D printing. The primary objective of this discussion is to apply these concepts and tools to the creation of prototype motors. At the end, the chapter includes examples to help readers practice their 2D design skills.

Chapter 5 delves into the extensive array of 3D operations offered by the Computer Assisted Design (CAD) software, Fusion 360. This chapter serves as a practical guide, enhancing user understanding through a variety of examples, tips, and insightful information related to each operation. These additions are specifically designed to facilitate the comprehension of the software's capabilities and to assist users in effectively utilizing these tools for designing and printing future 3D motor models. To consolidate learning and demonstrate the practical application of these skills, the chapter concludes with a comprehensive poly-component example. This final section showcases how multiple operations discussed throughout the chapter can be integrated to assemble a complex 3D model, providing readers with a clear understanding of how to apply these techniques in real-world projects.

Chapter 6 considers the critical steps required to transition from a 3D model to the actual fabrication of a motor. This journey encompasses a thorough analysis of the various parameters that need to be addressed during the 3D printing process, including adjustments specific to the chosen printing technology and material considerations. The chapter explores how different materials can affect the final product, such as variations in strength, flexibility, and heat resistance, which are crucial for the functionality of the motor in its intended application. To aid in preparation and ensure a successful print, the chapter concludes with a concise list of ten essential elements to review before initiating the 3D printing process. This checklist serves as a vital resource to optimize the printing setup, ensuring that each print yields the best possible results and adheres to the required specifications.

Chapter 7 is dedicated to dissecting the process of motor design, beginning with a comprehensive definition of design and an overview of the various existing motors, along with the possible configurations of magnet arrangements. This is followed by a detailed examination of the components that make up a motor, including the calculation of selecting the optimal bearing and an exploration of the trigonometric entities that define the most critical diameter of a motor. Additionally, the chapter delves into the positional relationships of motors, the construction of various prototypes designed for academic purposes, and the in-depth technical explanation of the design of key components in the motors.

We wish to thank many people who were involved in writing this book. We express our gratitude to Charles B. Glaser who supported this book project.

CUCEI, Guadalajara, Mexico   Erik Cuevas
                                                                            Daniel Zaldivar
                                                                             Ernesto Ayala
                                                                           Óscar González
                                                                            Fernando Vega

Contents

Fundamentals of Electromagnetism

1

1.1   Magnetic Field

One of the principles by which it is possible to create a movement in electric motors is the generation of a magnetic field. This happens by circulating an electric current through a conductor. Once this field is produced and it is approached to another one (generally coming from a permanent magnet), an interaction will be created that will result in a force. In Fig. 1.1 it is possible to observe a representation of this phenomenon [1].

To take full advantage of this phenomenon and to produce a greater force, these magnetic fields should be as large as possible, this is achieved on the one hand by increasing the size of the magnets and/or their magnetization, and on the other hand the size of the conductor is achieved by creating coils of the conductor in question (turns of wire). From the above, it can be concluded that the magnitude of the force will depend directly on the current in the conductor (coil) and the strength of the magnetic field (permanent magnet), and that the resulting force will increase when the magnetic field is perpendicular to the conductor.

Now that it is explained how the force is produced to move the rotor (moving part of the motor) of a motor, it will be interesting to know, what is the shape of a magnetic field? This shape can be seen in Fig. 1.2, which although it is an abstract idea allows us to generate a very convenient visual representation for demonstration purposes and is also corroborated with experiments such as bringing a magnet close to iron filings as in Fig. 1.3(a), or with devices such as the magnetic field viewer shown in Fig. 1.3(b).

These methods allows to visualize the patterns formed by the magnetic flux lines, and quickly quantify the magnetization of the magnets, or the strength of a coil to which an electric current is supplied, as well as to determine the effects that they will produce in the mechanisms that are coupled. As standar convention the direction of the flux lines has the orientation of the north pole to the south of the permanent magnet.

© The Author(s), under exclusive license to Springer Nature Switzerland AG 2025
E. Cuevas et al., *DC Motors*, Synthesis Lectures on Engineering, Science, and
Technology, https://doi.org/10.1007/978-3-031-64354-5_1

Fig. 1.1 Force resulting from
the approach of a magnet to a
conductor through which an
electric current is flowing

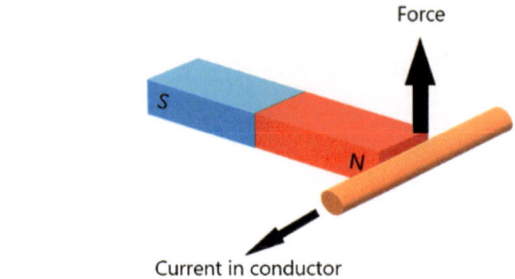

Fig. 1.2 Magnetic field lines
of a permanent magnet

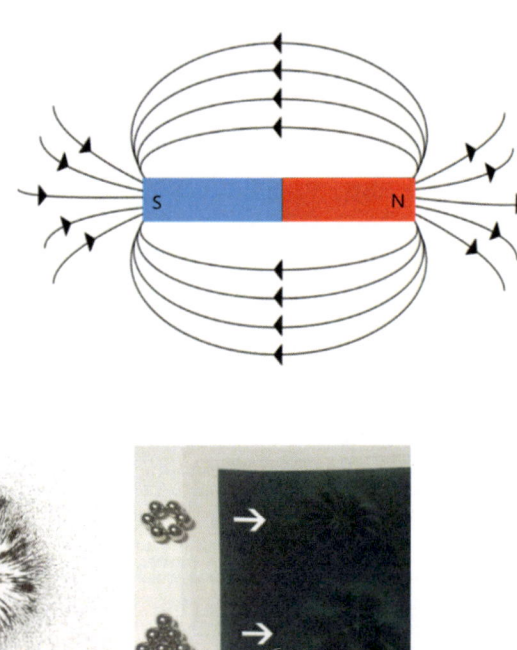

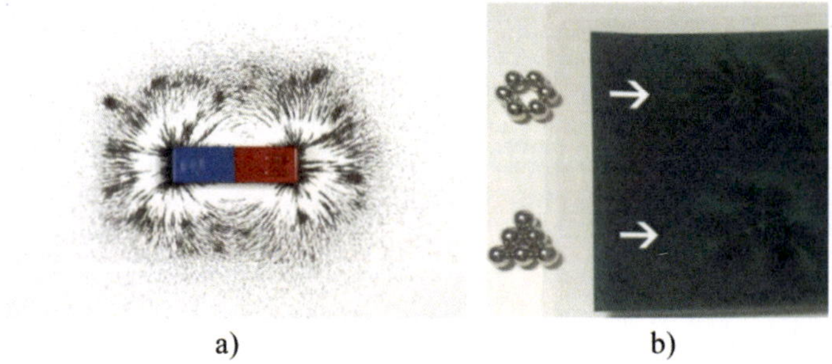

a)   b)

Fig. 1.3 (**a**) Magnetic field lines produced by iron filings, (**b**) magnetic field lines displayed with a
magnetic field viewer

1.2   Magnetic Flux Density

Now that we know the shape of the magnetic field in its 2-dimensional representation,
it must be considered that this field is present around a magnet or coil Fig. 1.4. This is
of vital importance, since the magnetic flux density is a measure of the magnetic field
strength and is defined as the magnetic flux density passing through a unit area, measured

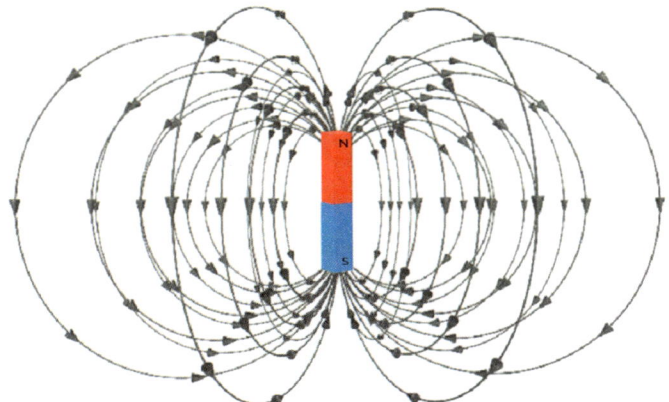

Fig. 1.4 3-D view of magnetic flux in a permanent magnet

perpendicular to the direction of the magnetic flux, which can be described in terms of a vector.

Therefore, the flux density is simply the magnetic flux divided by the cross-sectional area, as shown in Eq. 1.1:

$$B = \frac{\Phi}{A} \qquad (1.1)$$

The magnitude of the flux density is given by Eq. 1.1, and its direction is that of the predominant flux in the lines at each point. Near the top of the magnet in Fig. 1.4, for example, the flux density will be higher and in the direction toward the top of the magnet, while in the middle and away from the body of the magnet, the flux density will be lower and in the direction toward the bottom of the magnet.

As can be seen in Fig. 1.4, the magnetic flux is found surrounding the permanent magnet. This phenomenon is also applicable to a coil, since the magnetic field is not well oriented towards only one end, the coil must be built around a ferromagnetic core Fig. 1.5. This core is made of silicon iron sheets due to its high magnetic permeability (μ). Magnetic permeability is defined as the ability of conductors to affect and be affected by magnetic fields, as well as the ability to become sources of magnetic fields, i.e., the ability to create them without the need for external currents. The core, together with the coil (winding), is an essential component in many electrical and electronic devices. Its main application is in transformers, electric motors, relays and other power and signal applications. It should be clarified that the laminated silicon iron core is not the only option to take advantage of the electromagnetic field, since, as will be seen in later chapters, it is also possible to use composite materials, for example, the ferromagnetic PLA filament, which is useful to build prototypes of low-cost and fast manufacturing motors.

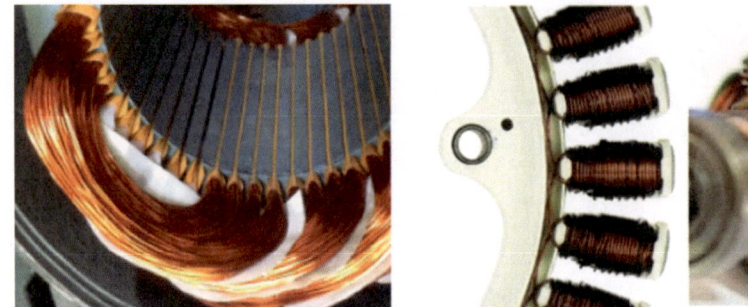

Fig. 1.5 Examples of motor windings with laminated silicon steel cores

Although its final efficiency is around 80%, it is quite acceptable before moving on to the manufacturing process of a final motor.

Thanks to these ferromagnetic materials that serve to form and direct magnetic fields, energy transfer and conversion can be achieved. Without these materials the practical implementation of most common electromechanical energy conversion devices would not be possible. The ability to analyze and describe systems employing these materials is essential to design and understand devices such as electric motors.

Finally, the units designated for magnetic field and magnetic flux in the international system of units are: for magnetic field, the Tesla (T), while the unit for magnetic flux is Tesla/m^2. This combination of units previously had the historical name of Weber (Wb). This change was in honor of Nikola Tesla, who is credited with the invention of the induction motor.

1.3   Ampere's Law

The physicist and mathematician André-Marie Ampere (1775–1836) enunciated one of the main theorems of electromagnetism, which models the relationship of a static magnetic field to the cause that produces it, i.e., a stationary electric current. James Clerk Maxwell later corrected it and it is now one of Maxwell's equations, forming part of the electromagnetism of classical physics [2]. Ampere's law determines that the circulation of the magnetic field along a closed line (surface), is equivalent to the algebraic sum of the intensities of the currents that cross the surface delimited by the closed line, multiplied by the permittivity of the medium, in other words and under the conditions mentioned above, the circulation of the intensity of the magnetic field in a closed contour is proportional to the current that crosses it. This is expressed by Eq. 1.2:

$$\oint_C \vec{B} \cdot \vec{dl} = \mu_0 \sum I \tag{1.2}$$

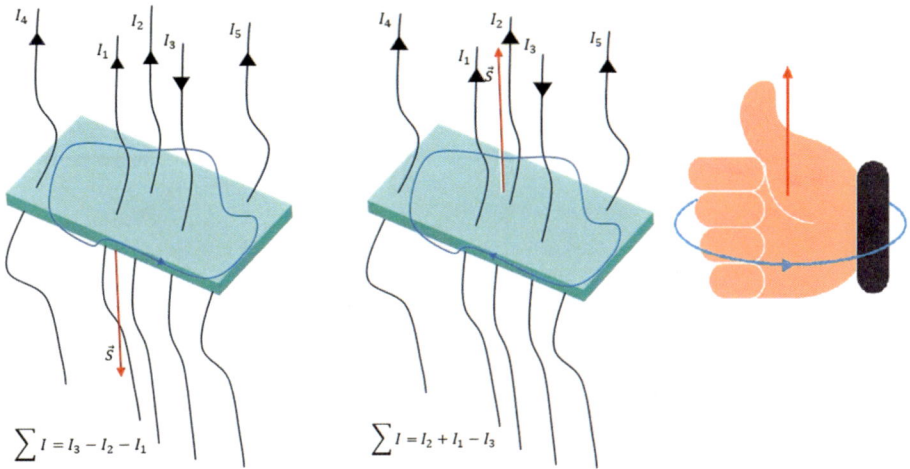

Fig. 1.6 Circulation of magnetic field strengths along a closed line

The intensities passing through the closed line can have different directions and consequently some will be considered positive and others negative. To determine the sign of these intensities, it is necessary to determine the surface vector formed by the closed line. In other words, if the direction of the intensities coincides with the direction of the surface vector, the intensity will be considered positive, therefore, if it is oriented in the opposite direction the intensity will be considered negative Fig. 1.6.

Some of the advantages that Ampere's law allows us to study the magnetic fields generated by electric currents are:

– The calculation of the magnetic field generated by electric currents when certain conditions occur and a suitable closed line is chosen.
– Because the magnetic field along a closed line is not zero, magnetic fields are not conserved and, therefore, there is no magnetic scalar potential.

1.4   Faraday's Law and Electromagnetic Induction

The energy coming from any source, medium or device that supplies electric current is known as electromotive force (e.m.f.) such force can be originated by the phenomenon known and discovered by Michael Faraday 1831 as electromagnetic induction. His discovery states that, if a variable magnetic field flux exists in a coil, then an e.m.f. will be induced in the coil. In other words, an electric current is established in the coil without an energy source being connected to it. From the above mentioned Faraday's law is established, which tells us that the e.m.f. induced in a circuit is related to the variation of the

Fig. 1.7 Potential difference
in a coil induced by the
variation of the magnetic field
(movement of a permanent
magnet)

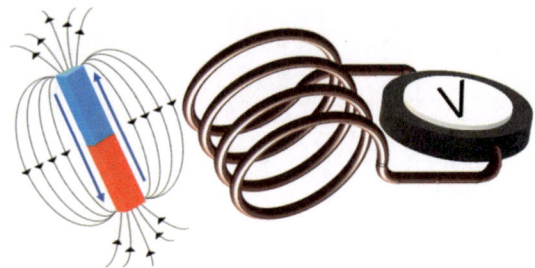

magnetic flux. This is expressed by Eq. 1.3:

$$\varepsilon = -\frac{d\Phi}{dt}$$ (1.3)

where the negative sign in Eq. 1.3 indicates that the direction of the induced e.m.f. is such
that it opposes the change that produces it, this is by the incorporation of the Lenz law
proposed by Heinrich Lenz in 1833 and which is collectively known as the Faraday-Lenz
law.

Since, in practice, not only one wire loop but a set of wire loops is used, where each
one contributes the same e.m.f., an additional term N must be added to represent the
number of turns. This is expressed by Eq. 1.4:

$$\varepsilon = -N\frac{d\Phi}{dt}$$ (1.4)

Because Michael Faraday's discovery is based on a relatively simple experiment, it can
be replicated as follows, using a cardboard tube with insulated wire wrapped around it to
form a coil. By connecting a voltmeter to the ends of the coil the induced e.m.f. can be
observed as a magnet is passed through the coil. The experiment is shown in Fig. 1.7.

1.5   Understanding Magnetic Forces in Motors

One of the basic principles of energy is that it is neither created nor destroyed, it is
simply transformed, and in the human quest to convert different types of energy into
work, electrical energy has become an efficient, cheap and in some cases clean form.

While the magnetic fields coming from permanent magnets or electromagnets may
be familiar, they seem to have meanings full of mystery. However, such magnetic fields
can be visible with different phenomena that arise through interaction with ferromagnetic
materials and even the effects of attraction and repulsion that occur between two fields of
different sign Fig. 1.8. In the nineteenth century there were quite interesting observations

Fig. 1.8 Physical motion of a
needle caused by the repulsion
of magnetic fields

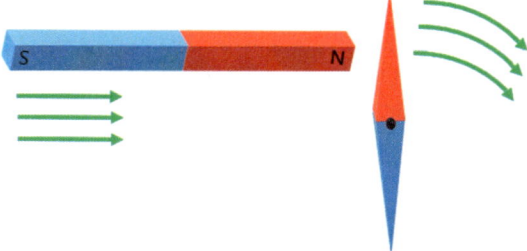

between moving electric charges and magnetism and how these could be reversible. However, it was not until the discovery of the physical motion produced by the interaction of magnetic fields that the first electric motors and generators were born.

The first device capable of converting rotating physical motion into electricity by means of magnetic fields was the Faraday disk Fig. 1.9. Although the discovery of the Faraday disk was functional, it was inefficient due to the counter-flows of current. This is because while a flux is induced directly under the magnet, the current flows in the opposite direction in regions outside the influence of the magnetic field. Because of this counter-flow the power was limited at the wire ends and caused overheating of the copper disk and consequently dissipation losses. Later homopolar motors solved this problem by using a series of magnets arranged around the perimeter of the disk in order to keep the field constant radially from the center to the edge, thus eliminating the areas where the backflow occurred.

Subsequently, a new DC motor appeared which contained mechanical switches in charge of synchronizing the current flow in the windings, managing to maintain a uni-directional torque under the influence of a magnetic field. This is the origin of a new commutation system that exists to this day and is called brush system. Thanks to this first DC motor it was possible to establish that the polarity of the DC motor source will determine the direction of rotation.

As shown in Fig. 1.10, when the coil is aligned with the permanent magnets is when the maximum torque is produced, on the contrary, when it is at 90° with respect to the magnets the torque will be zero, due to this arrangement with a single armature coil a

Fig. 1.9 Rotating floating
conductor disk in a magnetic
field with an electrical contact
near the shaft and another on
the perimeter (Faraday disk)

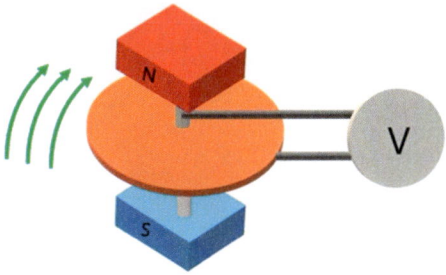

pulsating torque will be generated. To solve this pulsating effect, it should be consider placing multiple coils spaced so that one or more are always in a torque generating position Fig. 1.11, therefore, with a greater number of coils the movement will be smoother and continuous.

From the previous ideas of magnetic fields capable of producing rotating movements in a DC motor, a new revolutionary invention developed by Nikola Tesla and patented in 1888 called rotating magnetic field arises. This rotating magnetic field is a magnetic field that rotates at a uniform speed and is generated from an alternating electric current, hence the birth of alternating current (AC) motors. Rotating magnetic fields refer to magnetic fields that change direction or intensity in a rotating or cyclic pattern.

The operation of AC motors depends directly on the spatial arrangement of the windings, since by distributing on a ferromagnetic iron cylinder (stator) some coils, the inputs

Fig. 1.10 Example of a simple DC motor system with brushes

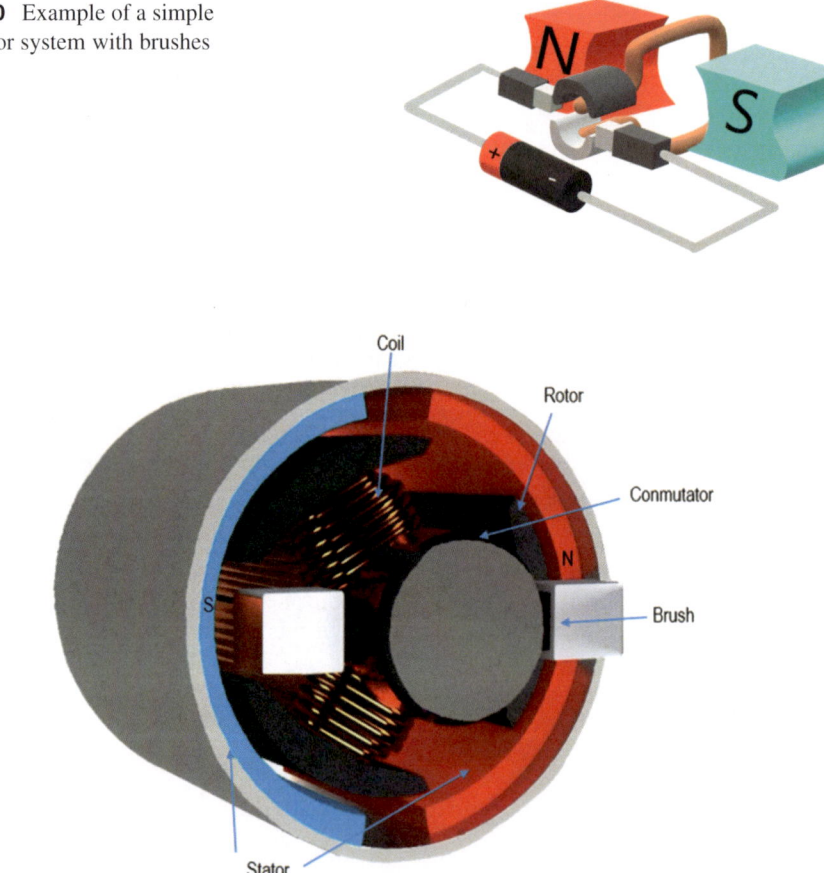

Fig. 1.11 Example of a DC motor with brushes and three coils

Table 1.1 Relationship between the number of poles and frequency of the source that determines the speed of rotation of the motor (RPM)

Poles	RPM at 50 Hz	RPM at 60 Hz
2	3000	3600
4	1500	1800
6	1000	1200
8	750	900
10	600	720
12	500	600
14	428.6	514.3
16	375	450
18	333.3	400
20	300	360

and outputs are separated 120° from each other and are fed with an alternating current, obtaining by the effect of the current conducted through them, a pulsating magnetic field. The number of times a winding appears in the motor will determine the number of poles, therefore, the speed of rotation of an alternating current motor will depend directly on the number of poles and the frequency of operation. Unlike DC motors, in AC motors if the speed of rotation is required to be modified, it is achieved by modifying the frequency of the power supply, an example of these speeds is shown in Table 1.1.

Within the alternating current motors there are synchronous and asynchronous motors, their main difference being the speed of the electromagnetic field:

– In the synchronous motor, the rotor and the rotating magnetic field of the stator rotate at the same speed, i.e. synchronized.
– In the asynchronous motor, the rotor rotates slightly slower than the rotating magnetic field in the stator, therefore, the magnetic field is always a few degrees ahead of the rotor.

1.6   Maxwell's Equations

It is possible to perform a correct analysis of magnetic circuits by means of Maxwell's equations, being this scientist who proposed to divide the electromagnetic field into electric and magnetic fields. In this way Maxwell's equations allow us to know how charges and magnets produce disturbances to the electromagnetic field, as well as how the field disturbs itself [3].

Since an electromagnet bases its operation on an electric field, it is important to remember the definition of the electric field or electric field strength $\left(\vec{E} \right)$, which refers to the interaction between two electrically charged bodies, this interaction can be of attraction or repulsion, being the electric flux defined as the number of lines of force that cross a defined surface, Eq. 1.5. On the other hand, that the magnetic field is a disturbance in space created by charges in motion and only affects charges that are also in motion.

$$\emptyset E = E \cdot dA \tag{1.5}$$

The first law of Maxwell's equations is *Gauss's law of the electric field* (Eq. 1.6), which states that the flow of electric fields through a closed surface is proportional to the magnitude of the sources of the electric field inside the same surface. Gauss equation (Eq. 1.6) describes how charges affect the electric field, and also states that the divergence of the electric field ($\vec{V} \cdot \vec{E}$), is directly proportional to the charge density. This divergence will be positive, if there are positive charges and negative, if there are negative charges. Due to the above, it is determined that charges of the same sign repel each other, while those of opposite sign attract each other. On the other hand, Gauss's law also establishes that the electric field will present losses, proportional to the distance, in the order of $\left(\frac{1}{r^2} \right)$.

$$\vec{V} \cdot \vec{E} = \frac{\rho}{\varepsilon_0} \tag{1.6}$$

The second *Gauss's law* is for the *magnetic field*. The magnetic flux is the relationship between the number of magnetic field lines $\vec{B}$ that cross a surface. As with the electric field, the total amount of magnetic flux that crosses a given surface is proportional to the magnetic charge, which is defined by Eq. 1.7.

$$\emptyset_m = \int \vec{B} \cdot d\vec{s} \tag{1.7}$$

The third law corresponds to *Faraday's law* (Eq. 1.8), which relates the rate of change of magnetic flux passing through a loop to the magnitude of the electromotive force induced in the loop. In other words, it describes how an electric current produces a magnetic field and, conversely, how a changing magnetic field generates an electric current in a conductor: This relationship is defined in Eq. 1.8.

$$\varepsilon = -\frac{d\Phi}{dt} \tag{1.8}$$

Finally, Maxwell's fourth equation is the *Apere-Maxwell law* Eq. 1.9, which states that the magnetic field passing through a closed path enclosing an electric current, that causes this magnetic field, is equal to the space permeability constant (μ) multiplied by the sum of two types of currents (intensities); the total current and another current called displacement current. In general terms, this law expresses that a magnetic field is generated by both a conduction current and a displacement current. This equation is composed of two parts:

The first part of the fourth Maxwell equation ($\vec{\nabla} \times \vec{B} = \mu_0 \vec{J}$) is the work of the French physicist André-Marie Ampere, where J is the electric current density. Because of this, the expression shows that the primary sources of the magnetic field are electric currents, thus showing the relationship of magnetic fields with electric currents. The second part of the equation added by the mathematician and scientist James Clerk Maxwell shows that a variable electric field produces a magnetic field.

$$\vec{\nabla} \times \vec{B} = \mu_0 \vec{J} + \mu_0 \varepsilon_0 \frac{\partial \vec{E}}{\partial \vec{t}} \tag{1.9}$$

In Sect. 1.2, we have seen how a variable magnetic field can induce an electric field. In Eq. 1.9, an analogy is observed. How a variable electric field produces a magnetic field. Thanks to these discoveries and set of formulas it is possible to take them to an endless number of applications in different areas of engineering.

1.7 Magnetic Materials

Magnetic materials are elements that exhibit magnetic properties, which means that they have the ability to generate a magnetic field. These materials will have different behaviors when faced with the action of an external magnetic field, which will lead to different values of absolute permeability (μ) that represents the influence of the magnetic properties of the environment on the induction. Additionally, and due to the different behaviors that magnetic fields have in different environments, the vacuum permeability (μ_0), is considered, based on the above, the relative magnetic permeability (μ_r) is established as the quotient between the absolute magnetic permeability and the vacuum magnetic permeability Eq. 1.10.

$$\mu_r = \mu / \mu_0 \tag{1.10}$$

Another property of magnetic materials is magnetic susceptibility (X_m), which is a physical property of a material that indicates the easiness with which a material can be magnetized. The properties described above make it possible to classify the magnetic behavior for materials according to the value of such parameters. Thus, materials can be classified into three main categories:

Ferromagnetic. These exhibit strong magnetic properties and are the most common. Some examples are: iron, nickel, cobalt, and some of their alloys. These materials have the property of retaining their magnetism after being exposed to an external magnetic field and are used in the manufacture of permanent magnets.

Diamagnetic. These are materials with a weak repulsion to the magnetic field, which means that their magnetization is opposite to the applied magnetic field. These materials

Table 1.2 Classification of magnetic materials

Material types	μ		μ_r		X_m
Diamagnetic	$\mu \approx \mu_0$	$\mu < \mu_0$	$\mu_r \approx 1$	$\mu_r < 1$	in the order of -10^{-5}
Paramagnetic		$\mu > \mu_0$		$\mu_r > 1$	in the order of 10^{-3}
Ferromagnetic	$\mu \gg \mu_0$		$\mu_r \gg 1$		Value within 10^3 y 10^5

experience a temporary magnetization in the opposite direction to the magnetic field, but their magnetic effect is quite weak compared to ferromagnetic and paramagnetic materials.

Paramagnetic. Although they possess weaker magnetic properties than ferromagnetic materials, these materials still respond to the presence of an external magnetic field. The atoms or molecules in these materials have magnetic moments that temporarily get aligned with the applied magnetic field, but lose their magnetism when the field is removed. In contrast to diamagnetism, diamagnetism does not depend on temperature, while paramagnetism is inversely proportional to temperature. As a result, the magnetic forces on the molecules are diminished when the agitation of the molecules increases with increasing temperature [4].

Table 1.2, shows the classification of diamagnetic, paramagnetic and ferromagnetic materials depending on their permeability and magnetic susceptibility.

Depending on their properties, magnetic materials have a wide range of applications, including the manufacture of magnets, cores and electronic devices. Their magnetic behavior will vary depending on their specific atomic composition and crystalline structure.

1.8   Magnetomotive Force (M.M.F.)

The magnetomotive force (m.m.f.) produces the magnetic flux and is determined by the reluctance, where the magnetic reluctance of a magnetic material or circuit is defined as the resistance it has to the passage of a magnetic flux when it is influenced by a magnetic field. In other words, it is the relationship between the magnetomotive force and the magnetic flux.

1.9   Analogy Between Magnetic and Electrical Circuits

The similarity that can be seen between these two types of circuits is that the magnetic flux and the electron flux are proportional to the magnetomotive and electromotive forces respectively as shown in Eqs. 1.11 and 1.12.

$$\text{Flux} = \frac{\text{m.m.f.}}{\text{Reluctance}} \tag{1.11}$$

$$\text{Current} = \frac{\text{e.m.f.}}{\text{Restitance}} \tag{1.12}$$

From the above is determined that to achieve an increase in the electric flux is by decreasing the reluctance, this is obtained by replacing the core of the winding (air) by a ferromagnetic material, for example, silicon steel, which achieves a higher magnetic flux in the circuit. Therefore, by confining the magnetic fields that were dispersed in the environment and forcing them to pass through a magnetic material with low reluctance, an increase in the magnetic flux is achieved, additionally the flux will depend on the current (I) flowing through the conductor and the number of turns (N) that it has. Using different techniques to manufacture air gaps (cores), different shapes and special characteristics can be obtained to operate different types of electrical machines.

1.10   Air Gaps

Another similarity that exists between these two types of circuits, is the easiness with which it is produced a flow either electrical or magnetic through the conductive medium (as long as this one has a low opposition to the flow). Due to this, in the electrical circuits electrical current will flow through the conductor as long as the circuit is closed. While in a magnetic circuit the flow is relatively easy to pass from a low reluctance material such as the air gap, into the air which weakens the magnetic fields. This is why, in the case of electric motors always try to minimize the space between its moving part (rotor) and static (stator), thus achieving a better use of the magnetic forces that will produce the work.

1.11   Saturation

Magnetic saturation is a property that refers to the precise point at which a magnetic material can no longer increase its magnetization in response to an external magnetic field. This is easily seen in Fig. 1.12 where the magnetization curve of a ferromagnetic material is shown. It is also noticeable how the permeability of paramagnetic materials is higher than that of vacuum and, on the contrary, the permeability of diamagnetic materials is lower. Specifically in the case of iron used in the cores of electric motors, it might be thinkable that the magnetic flux that circulates through this material is unlimited, but unfortunately this is not possible due to saturation.

When a magnetic material is exposed to a magnetic field, its magnetic dipoles tend to align with the direction of the field, and as the field strength increases, the magnetization

Fig. 1.12 Graph of the magnetic saturation and permeability curve of the materials

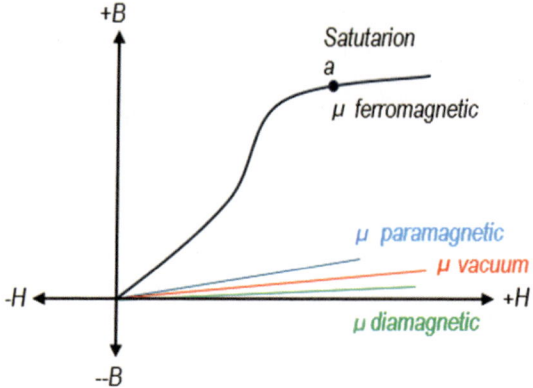

also increases. However, a point is reached at which almost all magnetic dipoles are aligned and therefore it is impossible to orient themselves any further. It is at this point that magnetic saturation reaches its limit and any further attempt to increase the magnetic field will not result in a significant increase in magnetization.

Due to the above, it is important to consider such saturation when working with magnetic materials, as it can affect the performance of magnetic devices, such as transformers, inductors and magnetic cores.

1.12   Magnetic Hysteresis

Magnetic hysteresis is the particular behavior of ferromagnetic materials when subjected to changes in their magnetic field. Specifically, this phenomenon occurs when magnetizing a ferromagnetic material and it does not follow exactly the same curve when increasing and decreasing the magnetic excitation (applied magnetic field), after it reaches a saturation point, Fig. 1.13. This is due to the fact that even with the magnetic excitation at zero there will remain a certain magnetization of the material called remanent induction, causing a delay or inertia in the magnetic response of the material. In order to reverse this remanent induction, it is necessary to apply a coercive force by means of an opposite excitation or an inversed polarity of the magnetic fields. Therefore, the hysteresis curve shows the relationship between magnetization (resulting magnetic field) and magnetic excitation as the latter varies. This can be defined as a hysteresis cycle or process of magnetization and demagnetization of the material in response to alternating changes in magnetic excitation [5].

Magnetic materials respond in different ways to the hysteresis phenomenon, but in general they can be categorized as follows:

Fig. 1.13 Hysteresis cycle and magnetic saturation

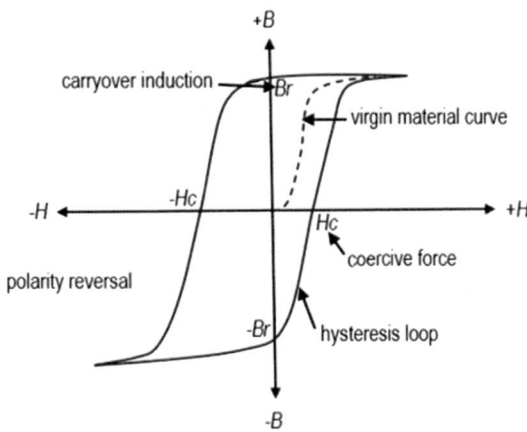

Hard magnetic materials: These materials possess a remanent induction and high coercive force. These types of materials are adequate for the manufacture of permanent magnets due to their resistance to demagnetization.

Soft magnetic materials: These materials have a low remanent induction and a small coercive force. They are used in applications such as transformers and electric motors.

An important consideration is hysteresis losses, which represent a quantity of energy that will be manifested in the form of heat in the air gap, making the device less efficient. In order to minimize these losses, the cores are made of magnetic materials such as silicon steel.

References

1. L. Alvarez, in *Electromagnetismo FMF-241* (Universidad Andrés Bello, 2014).
2. H. Young, R. Freedman, in *University Phisics*, 13th edn. (Pearson, 2011).
3. J. Reitz, F. Milford, R. Christy, in *Foundations of Electromagnetic Theory*, 4th edn. (Addison-Wesley, 2008).
4. M. Coey, in *Magnetism and Magnetic Materials*, 1st edn. (Cambridge University Press, 2010).
5. A. Hughes, in *Electric Motors and Drivers*, 3rd edn. (Elsevier, 2006).

Basic Concepts of Motors

2

2.1   Basic Concepts of Force, Power, and Torque

The torque generated by a brushless permanent magnet motor is arguably the most critical parameter to be assessed when evaluating motor performance [1]. This is because torque represents the turning or rotational force that the motor is capable of producing, essentially dictating the motor's ability to perform tasks that require rotational motion. As such, it stands as the primary specification that must be satisfied across a wide range of applications, regardless of the specific demands or operational context. Ensuring that a motor meets the required torque specifications is vital, as it directly impacts the efficiency, performance, and suitability of the motor for any given application.

In the area of physics, the concept of energy within a mechanical system is encapsulated by the term "work." Work is fundamentally defined as the product of a force applied to an object and the distance that object moves as a result of that force. To be more precise, work is calculated as the product of the displacement of the object and the component of the force that acts in the direction of this displacement. This relationship allows for a differential amount of mechanical energy to be expressed as $dW_m = Fdx$, where W_m denotes mechanical energy, F represents the force exerted in the x direction, dx signifies a differential length in the x direction and where v is the velocity of motion. Moving beyond this, power is defined as the quantity of work performed per unit of time, essentially measuring the rate at which energy is changed or transferred. Hence, based on the aforementioned description, mechanical power is delineated by what is referred to as Eq. 2.1, underscoring its derivation from the interplay between force, displacement, and the temporal dimension in which this interaction occurs.

$$P_m = \frac{dW_m}{dt} = F\frac{dx}{dt} = Fv \qquad (2.1)$$

© The Author(s), under exclusive license to Springer Nature Switzerland AG 2025   17
E. Cuevas et al., *DC Motors*, Synthesis Lectures on Engineering, Science, and
Technology, https://doi.org/10.1007/978-3-031-64354-5_2

When a tangential force F is applied at a radius r from the axis of rotation, it creates a torque T, which is determined by the product of the force F and the lever arm's length r, mathematically represented as $T = Fr$. This interaction underlines the principle that torque is essentially a rotational equivalent of linear force, acting over a distance to induce rotational motion. In this scenario, a differential amount of work done, can be calculated as $dW_m = Fdx = Frd\theta = Td\theta$, utilizing the relationship between the circumferential distance x and angular position θ, where $x = r\theta$. This equation elegantly links linear motion to rotational dynamics, illustrating how forces acting at a distance from a pivot point translate into rotational work. Subsequently, mechanical power, as previously defined, encapsulates the rate at which this work is performed over time, which is articulated in Eq. 2.2. This formulation underscores the fundamental principles of rotational dynamics, providing a quantitative measure of the energy transfer in systems subjected to rotational forces.

$$P_m = \frac{dW_m}{dt} = T\frac{d\theta}{dt} = T\omega \qquad (2.2)$$

where ω represents the rotational speed, evaluated in rad M/s.

Equations (2.1) and (2.2) serve as critical components in the engineering and design of motors, elucidating the relationships between torque, power, and the physical dimensions of the motor [2]. According to Eq. (2.1), torque is proportional to the square of the motor's diameter, and Eq. (2.2) establishes that power is directly proportional to torque. This implies that motors with larger diameters, where torque is generated, are capable of producing greater amounts of mechanical power, suggesting an initial design inclination towards maximizing motor diameter. However, this approach encounters several practical limitations. The most straightforward of these is the constraint imposed by the physical space available within a specific application, which may not accommodate a larger motor. Additionally, the volume of a motor increases as the square of its radius or diameter, leading to a scenario where simply enlarging the motor does not improve the ratio of output power to volume, due to the concurrent increase in mass—given that mass is proportional to volume. This ratio could only be improved if the average mass density of the motor's volume decreases as its diameter increases, which might be achievable by incorporating more air within the motor structure, thus enhancing power density marginally with increased diameter. The final constraint, inertia, becomes particularly relevant in scenarios demanding a high torque to inertia ratio. The inertia of a rotor scales with the fourth power of its radius or diameter, meaning the torque to inertia ratio decreases with the square of the rotor's radius. This reveals a crucial trade-off in motor design: in applications where minimizing inertia is paramount, opting for a large diameter becomes counterproductive, highlighting the intricate balance engineers must strike between maximizing power output and adhering to the physical and operational constraints of the motor design.

Equations (2.1) and (2.2) not only highlight the role of diameter in motor design but also point towards additional factors that can enhance the power output of a motor, once its diameter has been established [3]. Firstly, increasing the operational speed of the motor presents a method to augment power. However, in many instances, the speed at which a load operates is predefined, necessitating the use of gearings or other forms of speed reduction mechanisms to bridge the motor's output with the load's requirements. While conceptually straightforward, introducing speed reduction components comes with its own set of challenges, including added volume, mass, inertia, costs, potential for increased frictional losses, and a decrease in overall reliability, all of which must be carefully balanced against the advantages of operating at a higher speed.

Alternatively, enhancing the power output of a motor with a fixed diameter can be achieved by increasing the force density on the rotor. This involves elevating the electrical and magnetic operating conditions of the motor, often referred to as electric and magnetic loadings, respectively. Elevating the electrical operating point typically means increasing the current flowing through the motor, which in turn introduces additional ohmic I^2R losses. These losses, which escalate with the square of the current, necessitate effective heat dissipation strategies to prevent overheating. On the magnetic side, boosting the operating point may require the use of more or higher-quality magnetic materials, or innovative motor designs that better concentrate the magnetic flux across the air gap. Such modifications, while effective in increasing force density, inevitably lead to an increase in the motor's mass and volume due to the additional magnet and ferromagnetic materials required for flux concentration. These factors underscore the intricate balance between enhancing motor performance and managing the resulting increases in physical and operational constraints.

The complexities involved in optimizing motor performance while minimizing cost, as illuminated by the preceding discussion, underscore the fact that achieving high performance from a motor is anything but straightforward. When the objective is to extract the maximum possible performance, the physical limitations of the motor design are invariably stretched to their utmost capacities. This necessitates a comprehensive approach to high performance motor design, wherein every conceivable physical constraint must be acknowledged and integrated into the design phase. Engineering practices caution that omitting any constraint from the design process can lead to scenarios where those overlooked factors are strained beyond viable limits, potentially compromising the motor's functionality, durability, and cost-effectiveness. Therefore, a meticulous and holistic consideration of all constraints is critical to engineering motors that not only meet high performance standards but do so within realistic, practical boundaries.

2.2 Macroscopic Point of View of Torque

Determining the torque generated by a magnetic field within a motor can be approached from two distinct perspectives. The first method adopts a macroscopic viewpoint, heavily reliant on the principle of energy conservation [4]. This approach conceptualizes all

forms of losses associated with the motor—be they electrical, magnetic, or mechanical—
as external factors, thereby treating the system as conservative with no intrinsic energy
loss. Consequently, any addition of electrical energy is seen as being either stored within
the magnetic field or converted into mechanical output energy. This is encapsulated by the
equation $dW_e = dW + dW_m$, where dW_e, dW, and dW_m represent the differential quantities of electrical, magnetic, and mechanical energies, respectively. Under this framework,
it becomes feasible to establish a relationship between torque and the rate at which magnetic field energy changes, as delineated in Eq. 2.3. This equation underscores the inherent
link between the conservation of energy within the system and the mechanical torque output, illustrating how changes in the magnetic field's energy directly influence the torque
produced by the motor.

$$T = \left. \frac{-\partial W}{\partial \theta} \right|_{\lambda = \text{constant}} \tag{2.3}$$

This expression can be related to the rate of change as follows:

$$T = \left. \frac{-\partial W_c}{\partial \theta} \right|_{i = \text{constant}} \tag{2.4}$$

The foundational principles and mathematical derivations leading to Eqs. (2.3) and
(2.4) are well-documented across various technical and academic references, offering a
rich resource for those keen on delving deeper into the subject matter. As highlighted
earlier, the generation of positive torque is intricately linked to the dynamics of energy
within the motor, specifically how it serves to reduce stored energy under constant flux
conditions, and conversely, to augment energy storage under a regime of constant current.
While Eqs. (2.3) and (2.4) are applicable across a broad spectrum of scenarios, their application becomes particularly streamlined when addressing linear cases. Notably, Eq. (2.4)
emerges as the more pragmatic choice for practical computations due to its framing of
energy in terms of current rather than flux linkage. Despite the fact that both equations
ultimately converge on identical outcomes when equivalent substitutions are applied, this
discussion will primarily focus on the insights and results derived from Eq. (2.4) due to
its convenience and direct applicability to a wider range of motor design considerations.
The specific application of Eq. (2.4) in the context of mutual inductance is further elaborated upon in Eq. 2.5, demonstrating its utility in modeling and understanding the intricate
relationships between electrical currents, magnetic fields, and mechanical torque in motor
systems.

$$T = \frac{1}{2} i_1^2 \cdot \frac{dL_1}{d\theta} + \frac{1}{2} i_2^2 \cdot \frac{dL_2}{d\theta} + i_1 i_2 \cdot \frac{dL_{1,2}}{d\theta} \tag{2.5}$$

Delving into the nuances of the equation under discussion, particularly when isolating
its components for a more granular analysis, provides profound insights into the behavior
of electromagnetic systems. By initially considering a singly excited system, where the

secondary current, i^2 is set to zero, the focus narrows down to the equation's first term. This term elucidates that the torque generated within such a system is directly proportional to the square of the applied current, intriguingly indicating that the torque's magnitude is independent of the current's direction. This characteristic aligns with the operational principles of an electromagnet, which exerts an attractive force on a piece of steel regardless of the current flow direction. Further examination reveals that positive torque—or the force of attraction—is manifested whenever there's an increase in inductance. Essentially, this aspect of torque, commonly referred to as reluctance torque, perpetually strives to enhance inductance or permeance (given that inductance L is proportional to the square of the turns N and the permeance P), while concurrently reducing reluctance. The principle underlying this phenomenon can be intuitively understood through the electromagnet example, where the inductance and permeance of the magnetic circuit are augmented as the steel piece draws nearer to the electromagnet. This dynamic is a vivid demonstration of how electromagnetic systems leverage changes in inductance to generate torque, underpinning the significance of the equation's first term in elucidating the fundamental interactions at play within singly excited systems.

Shifting the focus from the initial assumption where the secondary current i_2 was considered to be zero, to a scenario where it is instead the primary current i_1 that equals zero, attention is directed towards the second term of Eq. 2.5. This adjustment in perspective reveals that the second term shares a fundamental similarity with the first, characterizing it also as reluctance torque, and thereby inheriting the same fundamental properties. This implies that, much like the first term, the torque produced under this condition is influenced by the square of the applied current, irrespective of its direction, reinforcing the concept that electromagnets exert force on steel regardless of the current flow direction, thus attributing to the action of reluctance torque.

When considering a situation where both self-inductances L_1 and L_2 are constant, the focus transitions exclusively to the last term in Eq. 2.5. This segment, differentiated by its dependence on the mutual flux or inductance between two coils, is referred to as the mutual torque or alignment torque. The existence and magnitude of this torque are contingent upon variations in the mutual inductance between the coils. Here, positive torque is generated when there is an increase in mutual coupling under conditions of congruent current signs, enhancing the alignment, and conversely, the torque acts to diminish mutual coupling when the currents bear opposite signs. This delineation between positive and negative torque—where the former arises from additive mutual fluxes and the latter from opposing ones—echoes the fundamental magnetic principle that opposite poles attract while like poles repel, offering a compelling insight into the interplay between electromagnetic forces within dual-coil systems.

Applying Eq. 2.4 to the specific scenario involving a mutually coupled coil system and a permanent magnet leads to a refined model, as articulated in Eq. 2.6. This progression in the equation's application signifies a pivotal step in understanding and quantifying the electromagnetic interactions within a system that integrates both electromagnetic coils and permanent magnetic materials. Equation 2.6, thus, encapsulates the dynamics of electromagnetic forces and torques arising from the synergy between an electrically induced magnetic field in the coils and the inherent magnetic field of the permanent magnet. This integrated model serves as a crucial analytical tool, providing a comprehensive framework for predicting the behavior of such a system under varying conditions. By factoring in the mutual coupling effects between the coil and the permanent magnet, Eq. 2.6 offers insights into the resultant torque and electromagnetic phenomena, thereby facilitating a deeper understanding of the operational characteristics and potential applications of electromechanical systems that harness both induced and permanent magnetic fields.

$$T = \frac{1}{2}i_1^2 \cdot \frac{dL}{d\theta} - \frac{1}{2}\theta^2 \cdot \frac{dR}{d\theta} + Ni\frac{d\phi}{d\theta} \tag{2.6}$$

Within the framework provided by the expression leading to Eq. 2.6, the first two terms are identified as the sources of reluctance torque originating from the coil and the magnet, respectively, while the third term represents the alignment torque that is the result of the mutual flux Φ linking the magnet to the coil. The architecture of this expression mirrors that of Eq. 2.5, with the first term maintaining an identical form, underscoring the consistent nature of reluctance torque across different electromagnetic configurations. The second term introduces a unique aspect of torque generation, being directly proportional to the square of the flux emanating from the magnet, independent of the flux's polarity. This term's negative sign stems from the inverse relationship between inductance and reluctance; as inductance decreases, reluctance increases, and vice versa, hence $dL/d\theta$ is proportional to $-dR/d\theta$. This relationship establishes a parallel between the first and second terms in terms of their contribution to torque production.

The connection to mutual flux linkage is further elaborated through the last term in Eq. 2.6, which draws a parallel with the equivalent term in Eq. 2.5. Here, the mutual flux linkage between the flux leaving the magnet ($\Phi\Phi$) and the coil is defined as $\lambda_m = N\phi$, mirroring the mutual inductance expression $\lambda_{12} = L_{12}i_2$ from Eq. 2.5. This correlation not only reinforces the fundamental principles governing electromagnetic interactions but also provides a comprehensive understanding of how mutual fluxes contribute to the overall torque production in systems involving both coils and permanent magnets, highlighting the nuanced interplay between electromagnetic fields and their mechanical manifestations.

In the context of a brushless permanent magnet motor, the dynamics articulated in Eq. 2.6 shed light on the sources of various torque components within the motor. The emergence of the first term is directly linked to the motor's construction, specifically when the winding inductance exhibits variability as a function of the rotor's position. This variability is a design characteristic that can lead to fluctuations in the magnetic field strength and, consequently, the torque produced. The second term in Eq. 2.6 encapsulates the concept of cogging torque, a phenomenon that arises when the flux from the magnets traverses paths of varying reluctance within the motor's magnetic circuit. This can lead to a degree of torque ripple or unwanted variations in the motor's output torque, affecting its smooth operation.

The final term in Eq. 2.6 is critical for the motor's function, as it represents the mutual torque—the fundamental force that drives the motor shaft to rotate. This torque results from the interaction of electromagnetic forces between the electromagnets in the stator and the permanent magnets attached to the rotor, manifesting through the cyclic attraction and repulsion between these components. It's this interplay that facilitates the conversion of electrical energy into mechanical motion, propelling the motor shaft. Among the three terms described in Eq. 2.6, the last term is inherently beneficial, embodying the desired mechanism for producing work. In contrast, the first two terms are typically regarded as parasitic effects—undesirable phenomena that detract from the motor's efficiency and performance. These effects introduce additional challenges in motor design, necessitating strategies to minimize their impact. Effective motor design, therefore, involves a delicate balance, optimizing the motor's architecture and materials to enhance the desired mutual torque while mitigating the effects of inductance variability and cogging to ensure smooth and efficient operation.

2.3   Microscopic Point of View of a Force

Exploring an alternative approach to understanding the forces at play within electromagnetic systems, one can derive an expression for mutual force grounded in the fundamental interaction between a moving point charge and a magnetic field [5]. This interaction is elegantly captured by the Lorentz force equation, $F = qv \times B$, where q represents the value of the charge, v its velocity, and B the magnetic flux density acting upon the charge. The cross-product symbol $\times$, signifies the vector cross product between the charge's velocity and the magnetic flux density. This relationship is crucial for calculating the direction and magnitude of the force exerted on the moving charge within a magnetic field. By applying the definition of the cross product, one can ascertain the magnitude of this force, a calculation formally presented in what is referred to as Eq. 2.7. This equation not only quantifies the force but also emphasizes the perpendicular nature of the interaction between the charge's path and the magnetic field lines, laying the groundwork for a deeper understanding of electromagnetic force dynamics and their practical implications in fields such as motor design and magnetic field analysis.

$$F = qv \cdot \sin(\alpha) \tag{2.7}$$

In this context, α represents the angle between the vectors v and B. The orientation of the resulting force is depicted in Fig. 2.1 and can be determined by applying the right-hand rule, which states that if one positions their right hand such that the fingers curl from v to B, then the thumb, when extended, will point in the direction of the force F. This principle highlights that the force reaches its peak when the vectors v and B

Fig. 2.1 Illustration of the
Lorentz force model

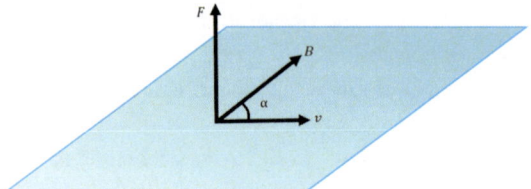

are orthogonal, a configuration often intentionally achieved in practical applications to maximize the force exerted. Consequently, for the purposes of subsequent analysis, it is standard to assume that α equals $\pi/2$, indicating a right angle between v and B.

Equation (2.7) finds its application in motor design through a series of further manipulations tailored to suit the context of electromagnetic forces in motion. By considering a differential charge element, denoted as dq, and redefining velocity in terms of distance over time—specifically as dl/dt, where l symbolizes the length along the path of motion—a refined expression of Eq. (2.7) is achieved. This adaptation allows for the representation of the force as a differential quantity, corresponding to the force exerted by the differential charge element in motion. The result of these manipulations is encapsulated in Eq. (2.8), which effectively describes the differential force generated by the moving differential charge within the magnetic field, thereby providing a more granular and applicable understanding of electromagnetic force dynamics relevant to motor design and analysis.

$$dF = dq\frac{dl}{dt}B \qquad (2.8)$$

Building upon the refined understanding of electromagnetic force in motor design, the next step involves acknowledging that current (i) can be defined as the time rate of change of charge, expressed mathematically as $i = dq/dt$. This realization allows for the transformation of the previously derived expression into $dF = iBdl$, signifying the differential force acting on a differential element of the wire due to the interaction with a magnetic field. This formulation directly links the magnitude of the current flowing through the wire, the strength of the magnetic field (B), and the length element (dl) of the wire within the magnetic field, to the force exerted. To calculate the total force experienced by a wire conducting a current i within a magnetic field B, one must integrate this expression over the length of the wire exposed to the field. The integration yields the total force, which is comprehensively presented in what is designated as Eq. 2.9. This equation not only encapsulates the cumulative effect of the magnetic field on the wire but also serves as a fundamental principle in designing and analyzing the interaction between current-carrying conductors and magnetic fields in various applications, including electric motors.

$$F = \int dF = \int iBdl \qquad (2.9)$$

When the wire is straight and encounters a uniform magnetic field across its entire length L, the solution to this integral is depicted in what is identified as Eq. 2.9. This scenario simplifies the calculation, as both the direction of the magnetic field and the magnitude of the force remain constant along the length of the wire, allowing for a straightforward integration over the specified distance L. This leads to the explicit expression captured in Eq. 2.10, which quantifies the total force exerted on the wire by the magnetic field, underlining the direct relationship between the physical dimensions of the wire, the characteristics of the magnetic field, and the resulting electromagnetic force.

$$F = BLi \tag{2.10}$$

The principle encapsulated by Eq. 2.10, often referred to as the BLi law, is instrumental for determining the force or torque generated by the interplay between a magnetic field and a wire carrying electrical current. Notably, Eq. 2.10 delineates a mutual force component that is remarkable for its independence from the magnetic field generated by the current i itself—a phenomenon traditionally known as the armature reaction. The overall magnetic field surrounding the wire thus emerges from the combination of the external magnetic field B and the field induced by the armature reaction, essentially a superposition of these two magnetic influences.

However, the introduction of nonlinear magnetic materials within the vicinity of these interacting fields can disrupt this straightforward superposition, making the resultant force or torque also contingent upon the armature reaction field. Particularly in scenarios where ferromagnetic materials are present—which is common in the regions of field interaction within motor designs—there arises a need to account for a potential diminution in the generated force attributable to armature reaction effects. This consideration underscores the complexity of accurately predicting motor performance and highlights the necessity of incorporating such nonlinear interactions into the motor design process to ensure optimal operational efficacy.

2.4   Mutual Torque and Reluctance

The discussions thus far illuminate that torque generation within motors can be attributed to two primary mechanisms, differentiated by their dependence on the variability of inductance with respect to the rotor's position [6]. Reluctance torque arises when there is a positional variation in self-inductance, whereas mutual or alignment torque is the product of changes in mutual inductance due to positional alterations. In practice, motor designs typically capitalize on one of these mechanisms for torque production. For instance, induction motors, DC brush and brushless motors, and synchronous motors are engineered to harness mutual torque. Conversely, switched reluctance motors are specifically designed to exploit reluctance torque for their operation.

In the context of motors that are optimized for mutual torque production, any torque emanating from fluctuations in self-inductance—referred to as reluctance torque—is generally considered parasitic, or undesirable. A common manifestation of this undesired torque is cogging torque, primarily induced by the presence of slots in the stator or rotor. This cogging effect introduces a ripple in the motor's output torque, leading to less smooth operation. Beyond cogging, other sources of reluctance torque can emerge from mechanical inaccuracies such as eccentricities and variations in dimensions, inherent to the manufacturing process. Given the preference for a constant torque output to ensure smooth mechanical motion, minimizing these parasitic torque components becomes a critical aspect of motor design, underscoring the importance of precision and innovation in the engineering of electric motors to achieve optimal performance.

To elucidate the process of calculating force within electromagnetic systems, let's examine the scenario depicted in Fig. 2.2. This setup features a voltage source that drives a current through a sliding bar. This current then interacts with a magnetic field that is oriented perpendicular to the plane of the paper, effectively directed into it. This configuration aligns with the principles discussed previously, where the motion of the sliding bar through the magnetic field generates a back electromotive force (EMF), denoted as e_b, across the conducting bars. This back EMF is a consequence of the electromagnetic induction occurring as the bar moves through the magnetic field, highlighting the fundamental interaction between electrical currents and magnetic fields that underpins the generation of force in such systems.

In determining the force exerted on the sliding bar as depicted in Fig. 2.2, both macroscopic and microscopic approaches, as previously delineated, are applicable. The microscopic method offers a direct pathway through the application of the BLi law, encapsulated in Eq. 2.10, which accurately calculates the force acting upon the sliding bar. By

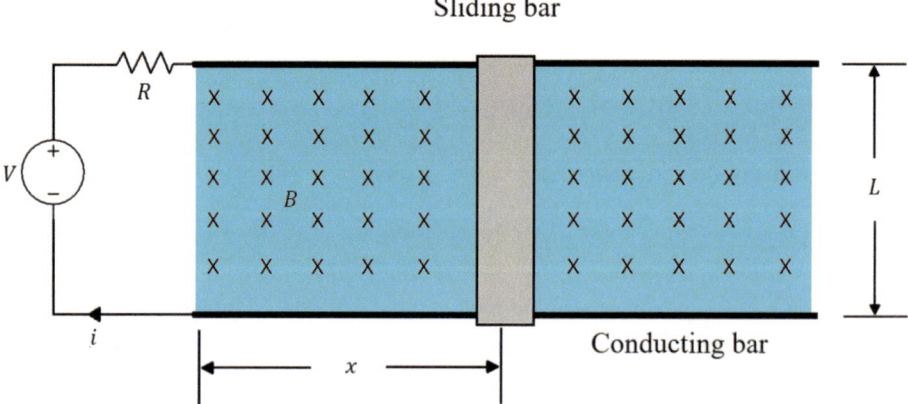

Fig. 2.2 A conceptual operation of a linear motor

employing the right-hand rule—a technique where the direction of motion is inferred by the orientation of the current (I) and the magnetic field (B)—it is established that the resultant motion of the bar is towards the right. Alternatively, this directional outcome can also be deduced by considering that the force naturally gravitates towards regions where the externally applied magnetic field B and the magnetic field generated by the current flowing through the bar counteract each other, or where the resultant magnetic field intensity is diminished. Observing the setup illustrated in Fig. 2.2, it becomes apparent that the magnetic fields are in opposition on the right side of the bar, leading to the conclusion that the force—and consequently, the motion of the bar—is directed to the right. This analysis not only corroborates the force's directionality but also underscores the utility of understanding magnetic field interactions for predicting motion in electromagnetic systems.

Utilizing the macroscopic energy method, as encapsulated in Eq. 2.6, offers a pertinent framework for analyzing the scenario presented in Fig. 2.2. This equation, particularly its last term, details the mutual torque that arises from the interaction between a magnetic flux and a current-carrying coil. When translating the principles of rotational motion and torque into the context of linear motion—as is the case with the sliding bar in Fig. 2.2—this last term can be adeptly reformulated to address the specifics of linear dynamics. The adaptation of this term for linear applications is meticulously illustrated in Eq. 2.11. This modification enables the equation to accurately represent the force developed through the interplay between the linearly moving bar and the magnetic flux, thus bridging the conceptual gap between rotational and linear electromagnetic interactions. Through this approach, Eq. 2.11 emerges as a crucial element in quantifying the linear force dynamics at play, underscoring the versatility of the macroscopic energy method in analyzing a wide range of electromagnetic phenomena.

$$F = Ni\frac{d\phi}{dx} \qquad (2.11)$$

In this context, ϕ represents the flux that interacts with the coil, which, for the purposes of this discussion, consists of a single turn. Referring to the arrangement illustrated in Fig. 2.2, the number of turns in the coil (N) is equal to 1, and the flux linkage ϕ can be expressed as BLx, where B is the magnetic field strength, L is the length of the bar, and x is the distance moved in the magnetic field. By integrating these specific values into the equation referenced as Eq. 2.10, the force exerted on the bar can once more be accurately represented by the BLi law, reaffirming the law's applicability and consistency in determining electromagnetic force based on the interplay of magnetic field strength, current, and physical dimensions of the conducting element within the field.

The configuration depicted in Fig. 2.2 exemplifies the dual functionality of such structures, capable of operating either as a motor or a generator, adhering to the principles of both the BLi and BLv laws concurrently. When an external force propels the bar in such a manner that the back electromotive force (e_b) surpasses the applied voltage (V), the current reverses direction, flowing back into the voltage source and consequently causing it to absorb power, thereby functioning as a generator. Conversely, in the absence of an externally applied force, the current drawn from the voltage source initiates movement of the sliding bar in the manner previously outlined, effectively utilizing the structure as a motor. This duality illustrates the inherent versatility of electromagnetic systems, which can alternate between converting mechanical energy to electrical energy and vice versa, based on the dynamics of current flow and the application of external forces, all within the framework of the fundamental *BLi* and *BLv* laws.

The application discussed offers a crucial insight into the interplay between electrical and mechanical power within an electromagnetic system, particularly evident when calculating the power dynamics of the setup involving the sliding bar. The electrical power (P_e) delivered to the sliding bar is determined by the product of the back electromotive force (e_b) and the current (i), expressed as $P_e = e_b i = BLvi$, applying the principles outlined in Eq. 2.2. When this logic is extended to compute the mechanical power (P_m), represented as $P_m = Fv = BLiv$, it becomes clear that both the electrical and mechanical powers are quantitatively equal to $BLvi$. This equality signifies that the output mechanical power directly mirrors the input electrical power, a relationship that consistently applies across all instances of mutual torque or force generation. This fundamental correspondence between electrical and mechanical power proves instrumental in determining back EMF, torque, or force within a system. Knowing any one of these quantities enables the calculation of the others. In scenarios involving rotational motion, this crucial relationship is further encapsulated by the formula presented in Eq. 2.12, highlighting the universal applicability of these principles in both linear and rotational electromagnetic systems.

$$e_b i = T \omega_m \tag{2.12}$$

where ω_m symbolizes the speed in *radM/s*.

Equation 2.12 is not only instrumental in the analysis of motor operations but also plays a crucial role in the design of motors. This equation indicates that, for a specified amount of mechanical output power, the necessary electrical input power can manifest in various forms: either through a high back electromotive force (EMF) paired with a low current, a high current coupled with a low back EMF, or an intermediate balance between the two. Among these options, designers often favor a configuration that achieves a high back EMF at a low current. This preference is rooted in the practical advantages such a design offers, notably in reducing the demand on the power electronics responsible for managing the motor's current. By opting for a high back EMF, the system can efficiently convert electrical power to mechanical power with minimal current, thus lessening the thermal and resistive losses typically associated with high current flows. This approach not only enhances the efficiency and performance of the motor but also contributes to the longevity and reliability of the power electronics, making it a strategic choice in motor design.

2.5   Fundamentals of Brushless Motors

Designing brushless permanent magnet motors involves a complex interplay of various scientific and engineering disciplines. At a broad level, proficiency in magnetics, mechanics, thermodynamics, electronics, acoustics, and material science is essential for crafting motors that meet diverse operational demands [7]. More specifically, understanding the performance requirements and the constraints of the intended application of the motor is crucial. This specialized knowledge allows designers to navigate through myriad design choices to find an optimal solution that balances performance with cost-efficiency. While this text primarily concentrates on the magnetic aspects of motor design, acknowledging their central role in functionality and efficiency, it is important to recognize that other aspects, though not detailed here, are also integral to the holistic design process. These include considerations of mechanical structure, heat management, electronic compatibility, noise levels, and material selection, each contributing to the overall effectiveness and reliability of the motor.

In addition to addressing the performance requirements essential for designing brushless permanent magnet motors, it is crucial to establish certain initial assumptions to sharpen and guide the early stages of the design process. These assumptions play a pivotal role, with some imposing specific restrictions while others point towards conventional design methodologies typically employed in the industry. A key assumption is that the motor is intended to produce rotary motion. Although the design principles discussed can also be adapted to motors designed for linear motion, the primary focus initially is on rotary motors where the rotor is situated within the stator. This focus helps streamline the design process by providing a clear and specific framework within which most conventional motor designs operate, thereby facilitating a more directed and efficient approach to developing motor solutions that meet predefined criteria and application-specific needs.

Most brushless permanent magnet motors feature magnets that are mounted on the rotor's surface, directly facing an air gap, making this configuration the primary focus of initial design efforts. This topology is commonly chosen due to its straightforward design and effective magnetic interaction between the rotor and the stator. However, an alternative approach involves embedding the permanent magnets within the rotor's steel structure, known as interior permanent magnet (IPM) topologies, which are favored for several reasons. Firstly, burying the magnets allows for flux concentration, enhancing the magnetic field's strength and efficiency. Secondly, enclosing the magnets within steel not only fortifies the rotor structurally but also enables the motor to operate at higher speeds due to increased durability and stability. Lastly, IPM motors can operate across a broader speed range. This flexibility is facilitated by field weakening control, which adjusts the motor's magnetic field to allow higher speeds beyond the base speed provided by the permanent magnets alone. These design considerations underscore the adaptability and technical considerations required to optimize motor performance for specific applications.

2.6   Main Concepts

In the area of brushless motor design, mutual torque and back electromotive force (EMF) stand out as critical parameters that dictate the motor's performance. These parameters are intricately connected, a relationship formalized by Eq. 2.11, which establishes that understanding one of these variables inherently provides insights into the other. While the BLi and BLv laws offer methods to measure torque and back EMF respectively, a more streamlined approach involves first calculating the flux linkage. Once the flux linkage is determined, Faraday's law can be applied to compute the back EMF accurately. Subsequently, employing Eq. 2.11 with this derived value allows for a precise calculation of the torque. This methodical approach not only simplifies the design process but also enhances the accuracy of predicting motor behavior under various operating conditions, making it a fundamental strategy in the development of efficient brushless motors.

The cross-sectional view of a motor, as illustrated in Fig. 2.3, presents a rotor equipped with $N_m = 4$ magnet poles facing the air gap. This setup introduces a factor of two difference between the electrical angle (θ_e) and the mechanical angle (θ_m), with the relationship defined as $\theta_e = (N_m/2)\theta_m$. This simplification facilitates the understanding of angular measurements in the motor's design. The depiction of the stator in this figure intentionally omits slots or windings to maintain clarity in illustrating the magnetic flux behavior.

In this motor configuration, the magnetic flux emanating from the North poles at the air gap diverges and crosses over to the stator. Upon reaching the stator, the flux splits into two equal parts, each flowing in opposite directions, before crossing the air gap again towards the South poles. The specific path of this flux, particularly for one half of a North

Fig. 2.3 Motor structure with its flux paths

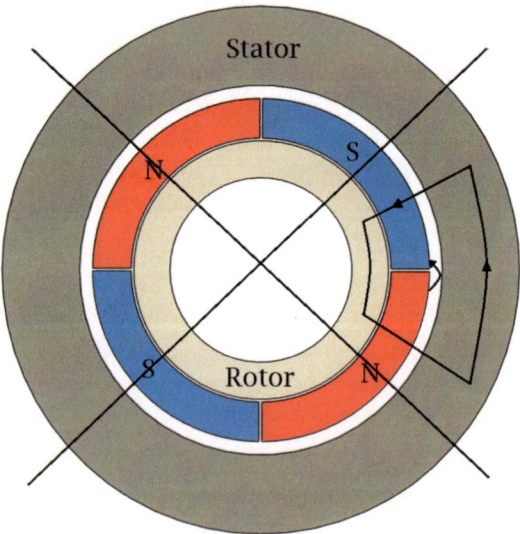

Fig. 2.4 Model of the magnetic circuit

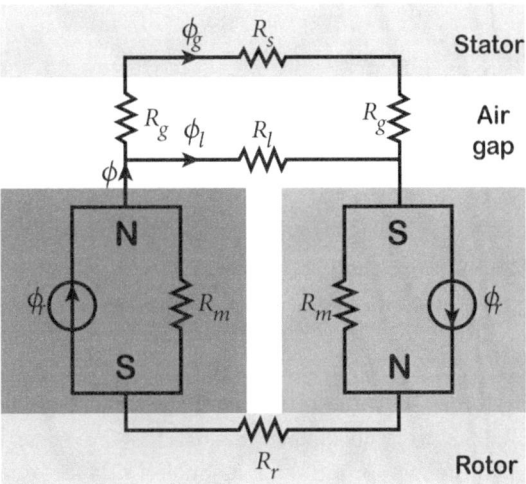

and South pole facing the air gap, is demonstrated on the right side of the figure. The flux dynamics for the remaining adjacent half-pole pairs mirror this pattern. Such a detailed portrayal helps in understanding the flux circulation within the motor, highlighting how magnetic interactions are orchestrated to achieve efficient motor function.

In the motor cross-section depicted in Fig. 2.3, while the primary flux path involves the transition of magnet flux from the North poles at the air gap into the stator and then back to the South poles, there is also an alternate path where some of the magnet flux directly jumps between adjacent magnets across the air gap. This secondary path of flux, visible on the right side of the figure, is referred to as magnet leakage flux. Given that the flux behavior shown in Fig. 2.3 is representative and repeats for each adjacent half pole pair, modeling efforts can be concentrated on just one pair as illustrated in the subsequent Fig. 2.4.

In this model, the rotor and stator are simplified into their respective reluctances, R_r for the rotor and R_s for the stator. The two halves of the magnets are represented as a flux source along with an associated magnet reluctance R_m, where the orientation of the flux source indicates the magnet polarity. The primary flux path from the magnets across the air gap into the stator traverses through the air gap reluctances labeled R_g. Meanwhile, the leakage flux, which bypasses the stator by moving directly from one magnet to the next, passes through a leakage reluctance R_l. This model therefore incorporates three distinct circuit fluxes: the magnet flux ϕ, the air gap flux ϕ_g, and the leakage flux ϕ_l, providing a structured and simplified approach to understanding the magnetic interactions within the motor.

Prior to calculating the back electromotive force (EMF), it is necessary to first solve for the air gap flux density B_g by analyzing the magnetic circuit. While a direct solution using the configuration in Fig. 2.4 is possible, it is more practical to simplify the

circuit as demonstrated in Fig. 2.5. In this revised model, the magnet on the right and
the rotor reluctance, which are originally in series, are repositioned (as seen in Fig. 2.5c)
to streamline the circuit. This adjustment groups the two half magnets together, aligning
them adjacent to one another, and relocates the rotor reluctance alongside the other circuit
reluctances.

Determining an analytical description of the leakage reluctance proves challenging;
however, it is feasible to estimate the proportion of the flux that traverses the primary flux
path across the air gap relative to the total magnet flux. This relationship can be expressed
as $\phi_g = K_l\phi$, where K_l is a leakage factor typically just under one, indicating most of
the flux follows the primary path. Consequently, to further simplify the magnetic circuit,
the leakage reluctance R_l can be disregarded, as depicted in Fig. 2.5b. This elimination
is justified because only a minimal amount of flux follows the leakage path, and it is
complicated to accurately quantify R_l.

To account for the small amount of flux that does follow the leakage path, the solution
for ϕ will be adjusted by multiplying it by the estimated K_l to approximate ϕ_g. With the
leakage path removed from consideration, the reluctances of the rotor and stator steel,

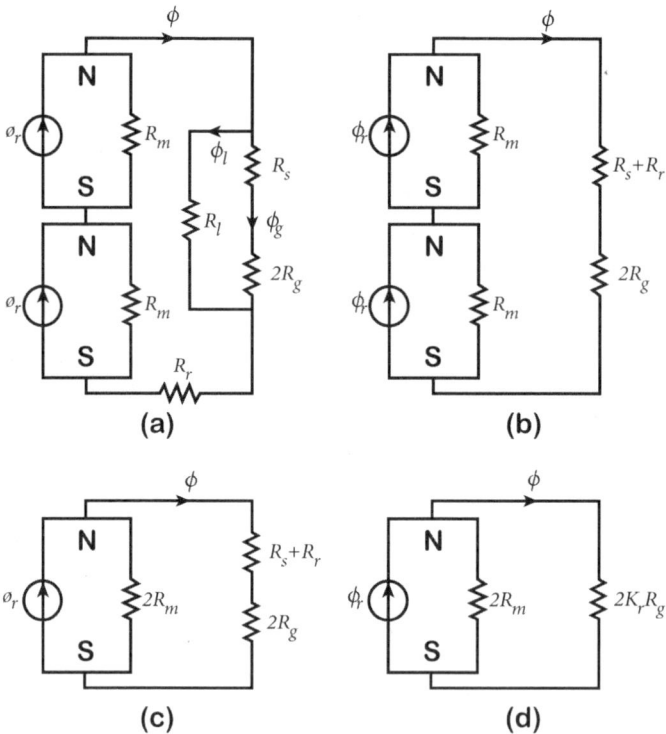

Fig. 2.5 Circuits which simplify the main circuit of Fig. 2.4

now effectively in series, can be combined into a single equivalent reluctance, as shown in the simplified diagram in Fig. 2.5b. This consolidation further refines the magnetic circuit, making it easier to analyze and solve for the crucial air gap flux density needed to determine the back EMF.

In Fig. 2.5b, the arrangement of the two magnet halves in series is further streamlined as depicted in Fig. 2.5c. This simplification can be analogized to an electrical circuit through the use of the Norton equivalent circuit method, which is typically employed to simplify networks of resistors and sources. By applying this method to the magnetic circuit, the combined effect of the two series magnet halves results in a simplified flux source (ϕ_r), which represents the flux that would be present if a hypothetical short were introduced across the series magnets. Additionally, the equivalent reluctance of this simplified setup is $2R_m$, reflecting the combined reluctance of the two magnet halves in series.

From a magnetic materials perspective, treating the two half magnets in series as a single unit equates to conceptualizing them as one block of permanent magnet material with twice the length of a single half. This conceptualization naturally leads to a doubling of the reluctance (R_m) because reluctance in a magnetic circuit is directly proportional to the length of the material through which the magnetic flux passes. Despite these modifications to the magnetic circuit's configuration, the total flux (ϕ_r) produced by the magnet remains unchanged. This unified approach not only simplifies the magnetic model but also helps in accurately assessing the magnet's contribution to the overall magnetic circuit without altering the flux output.

In the configuration presented in Fig. 2.5c, the combined steel reluctance, represented as $R_r + R_s$, exhibits nonlinear behavior due to the saturation characteristics inherent to ferromagnetic materials. To derive an analytical solution, it becomes necessary to address this complexity. Given that the permeability of steel is typically much higher relative to that of air, the steel reluctance tends to be considerably smaller than the air gap reluctance R_g. Under such conditions, the steel reluctance can be conceptualized as a minor perturbation to the air gap reluctance.

In practical terms, this means that the steel reluctance can effectively be disregarded by modifying the air gap reluctance slightly. This adjustment is achieved by introducing a reluctance factor K_r in the model, as illustrated in Fig. 2.5d. K_r is a constant that is slightly greater than one, designed to incrementally increase the air gap reluctance to account for the omitted steel reluctance. This adjustment helps simplify the model without significantly compromising its accuracy.

It is crucial to recognize that in real-world applications, deriving precise analytical expressions for the leakage factor K_l and the reluctance factor K_r is often impractical due to the simplifications made in such modeling efforts. These factors are typically determined based on empirical data and the designer's experience, reflecting the inherent challenges and limitations in accurately modeling complex magnetic interactions with basic theoretical constructs.

2.7   Solution for the Magnetic Circuit

Considering the magnetic circuit illustrated in Fig. 2.5d, the magnet flux ϕ can be articulated using the concept of flux division, similar to how current is divided between resistors in an electrical circuit [8]. This approach is mathematically formalized in Eq. 2.13. This equation models how the magnetic flux splits across different paths in the circuit, analogous to the distribution of electrical current through paths of varying resistance.

$$\phi = \frac{2R_m}{2R_m + 2K_r R_g}\phi_r = \frac{1}{1 + K_r \frac{R_g}{R_m}} \tag{2.13}$$

Drawing from the general formulas for magnet and air gap reluctances, the values for R_m and R_g can be calculated in the following manner:

$$R_m = \frac{l_m}{\mu_R \mu_O A_m}, R_g = \frac{g}{\mu_O A_g} \tag{2.14}$$

where the air gap flux can be modeled under the following expression:

$$\phi_g = K_l \phi = \frac{K_l}{1 + K_r \frac{\mu_R g A_m}{l_m A_g}} \tag{2.15}$$

In the context of determining magnetic reluctances, the variables l_m and A_m represent the length and cross-sectional area of the magnet, respectively, while g and A_g denote the length and cross-sectional area of the air gap. By introducing the flux concentration factor $C_\phi = A_m/A_g$, which quantifies the ratio of the magnet's cross-sectional area to that of the air gap, a more detailed analysis can be performed. Additionally, employing the flux density relationships $B_g = \phi_g/A_g$ for the air gap and $B_r = \phi_r/A_m$ for the magnet allows for a comprehensive understanding of how flux densities are distributed across different segments of the magnetic circuit. The permeance coefficient, expressed as $P_c = l_m/(gC_\phi)$, further refines this analysis by correlating the magnet's physical dimensions with the air gap and the flux concentration. Substituting these relationships into Eq. 2.14 results in a model that provides the air gap flux density as indicated by Eq. 2.16, thus capturing the nuanced interactions within the magnetic circuit and facilitating precise magnetic performance assessments.

$$B_g = \frac{K_l C_\phi}{1 + K_r \frac{\mu_R}{P_c}} \tag{2.16}$$

The equation in discussion quantifies the air gap flux density that traverses the air gap in a motor equipped with surface magnets. In such setups, the leakage factor (K_l) typically ranges between 0.9 and 1.0, the reluctance factor (K_r) between 1.0 and 1.2, and the flux concentration factor (C_ϕ) is ideally 1.0. These values, once set alongside the chosen magnet's remanence (B_r), influence the overall behavior of the air gap flux density, with the permeance coefficient (P_c) playing a crucial role in determining its amplitude. As the permeance coefficient increases, the air gap flux density nears its maximum potential, which is slightly less than the remanence of the magnet. Without flux concentration, achieving an air gap flux density (B_g) that surpasses B_r is unattainable. Additionally, the relationship between the permeance coefficient and the air gap flux density is inherently nonlinear, with the air gap flux density asymptotically approaching the remanence. This means that doubling P_c—which effectively doubles the length and volume of the magnet, thus significantly increasing both cost and material use—does not simply double B_g. Figure 2.6 illustrates the relationship between the permeance coefficient and the ratio B_g/B_r, with vertical lines indicating the typical permeance coefficient range of four to six, commonly utilized in many motor designs. This visual representation helps in understanding how variations in P_c affect motor performance and cost efficiency.

Equation 2.15 offers an approximation of the air gap flux density distributed across the surface of the magnet pole in a motor. Specifically, this equation calculates the magnitude of the air gap flux density $|B_g|$, as depicted in Fig. 2.7. It delineates how this magnitude is positive over North poles and negative over South poles, reflecting the directional properties of magnetic fields. Although this approximation may not perfectly capture all the nuances of the actual flux distribution, the formulation of Eq. 2.16 is highly informative for understanding motor dynamics. The equation not only sheds light on the operational aspects of motors but also emphasizes core principles that remain relevant and applicable in more precise and complex modeling scenarios. Thus, despite its simplicity, Eq. 2.16 is pivotal in illustrating fundamental electromagnetic behaviors crucial for both educational and practical applications in motor design.

In Fig. 2.7, for ease of representation and analysis, the horizontal axis is expressed in terms of electrical measure, which is periodic and corresponds to one pair of poles

Fig. 2.6 Relationship permanence coefficient an air gap flux

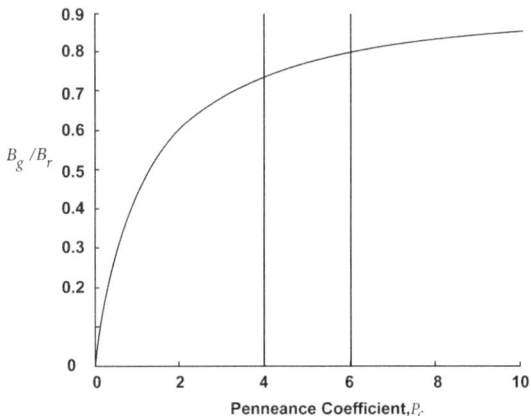

Fig. 2.7 Density distribution
of the ideal gap flux

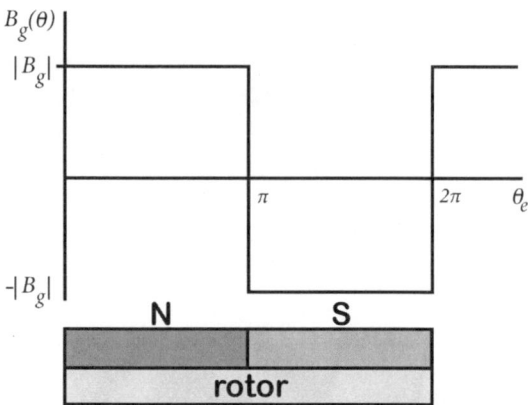

as illustrated in the diagram. This method of depiction allows for a clear and structured understanding of the magnetic field's variation over the motor's cycle. Specifically, in the case of the prototype motor under consideration, the rotor exhibits two electrical periods around its circumference. This means that the magnetic and electrical interactions repeat twice for each complete revolution of the rotor. By adopting this electrical measure, the depiction efficiently conveys the cyclical nature of the motor's electromagnetic behavior, facilitating a better grasp of how the flux density varies with position and how these variations influence motor performance.

2.8   Flux Linkage

With the magnetic circuit analysis completed, let's introduce an extension to the system: the incorporation of two slots that contain a winding made of N turns of wire, as depicted in Fig. 2.8. This arrangement forms a coil that exits the slot at the top of the figure and enters into the slot on the right side. The separation between the points where the coil enters and exits the slots corresponds to a coil pitch or coil throw of 180∘E or $\theta_p = 2\pi/N_m$radM, classified as a full pitch winding. This design ensures that the coil fully spans one complete cycle of the magnetic field across the rotor [9].

As the rotor rotates, the air gap flux interacts with this coil. Specifically, for the rotor position illustrated in Fig. 2.8, where $\theta_e = 0$, the flux flows toward the South pole of the rotor magnet and crosses the air gap from the coil's perspective. The direction of this flux flow is opposite to that which would be induced by current flowing through the coil, resulting in a negative flux linkage. Thus, if ϕ_g represents the air gap flux as specified in Eq. 2.14, the flux linkage at this rotor position is expressed as $\lambda = -N\phi_g$. This configuration highlights the dynamic interaction between the rotor's magnetic field and the coil, crucial for understanding how electrical output is influenced by mechanical rotation within the motor.

Fig. 2.8 Full-pitch coil for a
motor

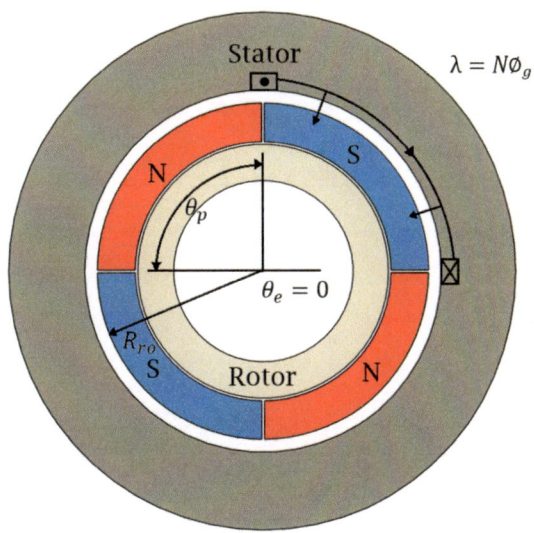

As the rotor rotates to a 90°E position, as illustrated in Fig. 2.9, the coil finds itself centered symmetrically over halves of both a South and a North pole. In this configuration, the flux direction differs over each pole: it flows toward the rotor over the South pole and away from the rotor over the North pole. Consequently, the magnetic fluxes linked by the coil from each half-pole counteract each other, resulting in a net flux linkage of zero. This scenario underscores the dynamic nature of magnetic interactions within the motor depending on rotor position.

Continuing the rotation by another 90°E brings the rotor to a position of 180°E, as depicted in Fig. 2.10. At this juncture, the coil is now directly centered over a North pole. The flux linkage at this position is equal in magnitude but opposite in direction to that at the $\theta_e = 0$ position, previously shown in another diagram. This reversal results in a positive flux linkage. These changes highlight how the rotational position of the rotor significantly influences the magnetic flux interactions within the coil, demonstrating a fundamental aspect of motor operation that is critical for understanding how electrical outputs are generated from mechanical motion.

As the rotor transitions from a 0°E position towards 180°E, the flux linkage experienced by the coil changes in a linear fashion, progressing from a minimum at 0°E to reach a maximum at 180°E. This linear variation occurs because the coil moves from aligning with a South pole to directly aligning with a North pole, causing an incremental increase in flux linkage due to the changing orientation of the magnetic field relative to the coil. Similarly, as the rotor continues its rotation past 180°E towards 360°E, the flux linkage undergoes a linear decline, returning from the maximum value at 180°E to another minimum when it completes the rotation at 360°E. This process describes a complete cycle of rotation where the flux linkage creates a periodic waveform, as exemplified in Fig. 2.11a.

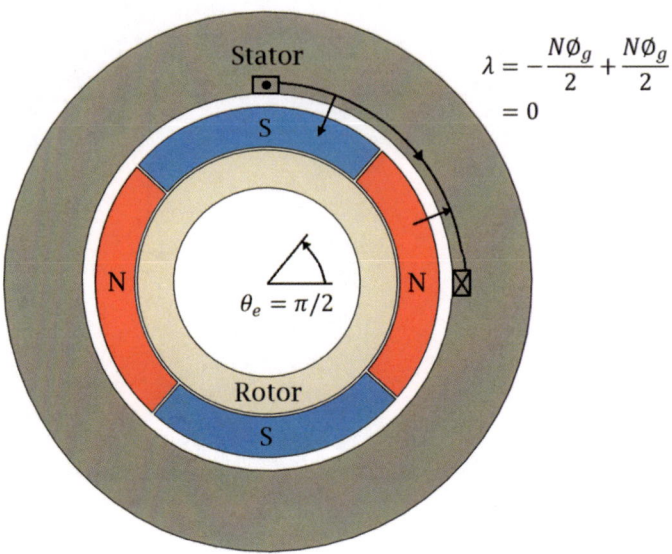

$$\lambda = -\frac{N\emptyset_g}{2} + \frac{N\emptyset_g}{2}$$
$$= 0$$

Fig. 2.9 Rotor at 90 for a motor

Fig. 2.10 Rotor at 180 for a
motor

$$\lambda = N\emptyset_g$$

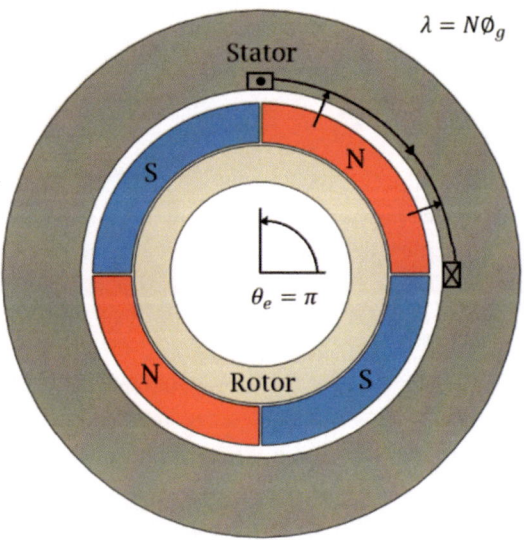

Each cycle of rotation leads to alternating linkage of South and North poles with the coil, illustrating the dynamic, cyclical nature of magnetic interactions within the motor, essential for generating electrical outputs based on mechanical motion.

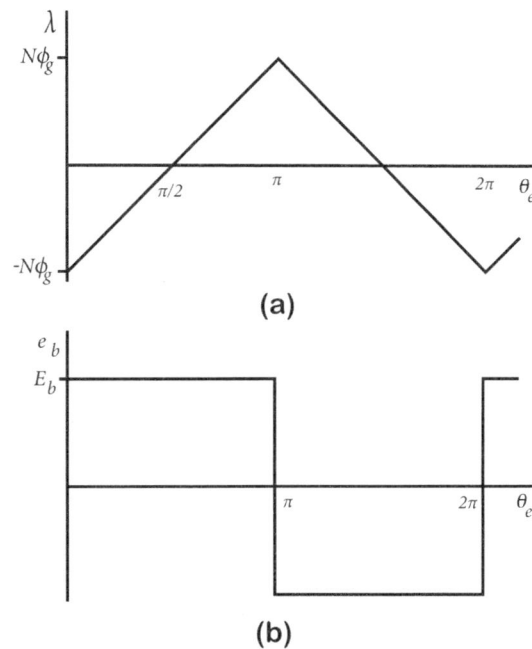

Fig. 2.11 Flux linkage
a waveform and **b** square wave

2.9   Torque and Back EMF

The flux linkage waveform depicted in Fig. 2.11a plays a crucial role in determining the characteristics of the back electromotive force (EMF) through Faraday's law, which states that the induced EMF in any closed circuit is equal to the rate of change of the magnetic flux through the circuit [10]. Given that the flux linkage waveform is triangular, following the rotor's cyclic motion between poles, the derivative of this waveform—necessary for calculating the back EMF—results in a square wave form. This is visually represented in Fig. 2.11b, where the back EMF waveform clearly contrasts with the triangular shape of the flux linkage due to its sharp transitions between maximum and minimum values.

The period of the back EMF waveform aligns perfectly with a 360°E rotation of the rotor, reinforcing the use of electrical measure as a standard metric in describing the electrical activity within the motor. This periodic nature of the back EMF is critical for synchronous operation and efficient electrical output in motor applications. Equation 2.17 provides an analytical expression for this back EMF, quantifying it in a form that can be readily applied in both theoretical analyses and practical engineering solutions to predict and enhance motor performance.

$$e_b = \frac{d\lambda}{dt} = \frac{d\theta_e}{dt} \cdot \frac{d\lambda}{d\theta_e} = \omega_e \frac{d\lambda}{d\theta_e} = \frac{N_m}{2}\omega_m \frac{d\lambda}{d\theta_e} = \frac{N_m}{2}\omega_m \frac{2N\phi_g}{\pi} \qquad (2.17)$$

where ω_m represents the rotational speed of the rotor in radians per second (radM/s). To further simplify the analysis, the air gap flux is detailed as presented in Eq. 2.18.

$$\phi_s = B_g A_g = B_g R_{r0} \theta_p L_{st} = \frac{2\pi}{N_m} B_g L_{st} R_{r0} \tag{2.18}$$

In this context, B_g represents the air gap flux density, θ_p is the angular pole pitch measured in radians per meter (radM), R_{r0} denotes the air gap radius at the surface where the magnets are mounted, and L_{st} refers to the axial length of the motor. These parameters together define the physical and magnetic properties of the air gap, crucial for determining the flux through the motor. By integrating this relationship into Eq. 2.19, which describes the back electromotive force (EMF), it is possible to calculate the amplitude of the back EMF (E_b). This step is essential for quantifying how changes in the motor's physical dimensions or its operational conditions directly influence the back EMF, ultimately affecting the motor's electrical output and efficiency.

$$E_b = |e_b| = \frac{N_m}{2} \omega_m \frac{2N}{\pi} \left(\frac{2\pi}{N_m} B_g L_{st} R_{r0} \right) = 2N B_g L_{st} R_{r0} \omega_m = K_e \omega_m \tag{2.19}$$

The expression for calculating the back electromotive force (EMF) aligns with the BLv law, which relates the velocity and magnetic flux to the induced EMF in a conductor. Specifically, the factor $2N$ in the expression accounts for the two slots in the motor, each containing N conductors. The term $R_{r0}\omega_m$ represents the linear velocity at which the magnetic flux linkage through the conductors changes as the rotor spins. This rate of change of flux linkage is crucial for determining the magnitude of the induced EMF. According to Eq. 2.19, all components of the expression, except for the composite term, contribute to a constant for the back EMF, denoted as K_e. This constant, K_e, essentially encapsulates the relationship between the back EMF and the rotor's angular velocity, with its units being volts per radian per meter per second (V/(radM/s)). This formulation simplifies the calculation of the back EMF by providing a straightforward multiplicative factor that can be applied to the motor's operational dynamics to estimate EMF output effectively.

The application of the models described helps in determining the torque generated when a current i flows through the coil of the motor. The model provides a simple algebraic relationship which correlates the torque with the position of the rotor. Given a constant current, the torque versus position curve mirrors the back EMF versus position curve depicted in Fig. 2.11. This similarity arises because both torque and back EMF are directly influenced by the flux linkage changes as the rotor moves. The magnitude or amplitude of the torque generated in this scenario is quantitatively defined by Eq. 2.20. This equation enables the precise calculation of torque based on current and rotor position, simplifying the evaluation of motor performance under various operational conditions.

$$|T| = \frac{E_b i}{\omega_m} = 2N B_g L_{st} R_{r0} i = K_t i \tag{2.20}$$

In the expression given by Eq. 2.20, all components except for R_{r0} define the force exerted on the rotor. When this force acts at the radius R_{r0}, it results in the generation of torque as per the relationship described in Eq. (2.1). From another analytical perspective, if you isolate the term i (current) in Eq. 2.20, the remaining factors together constitute a torque constant, denoted as K_t, with units of Newton meters per ampere (Nm/A). This torque constant K_t can be directly compared to the back EMF constant K_e previously discussed. Interestingly, these constants are revealed to represent the same fundamental quantity when analyzed in the context of their roles in motor performance, as highlighted in Eq. 2.21. This revelation underscores the inherent relationship between the electrical input (current) and mechanical output (torque) in the functioning of electric motors, illustrating how electrical properties and mechanical forces are interlinked within the motor's operational dynamics.

$$K_e = K_t = 2NB_gL_{st}R_{r0} \qquad (2.21)$$

The discussions of flux, flux linkage, back EMF, and torque within this section are based on idealized assumptions and simplifications. In practice, the characteristics of these variables can deviate significantly from the theoretical models. For instance, unlike the square wave representation of air gap flux density depicted in Fig. 2.9, actual flux densities often show more complex and varied patterns. Consequently, this discrepancy affects the flux linkage, which, instead of adopting the ideal triangular shape shown in Fig. 2.11a, might display irregular and less predictable shapes. Similarly, the back EMF, rather than forming a clean square wave as illustrated in Fig. 2.11b, often exhibits waveform distortions due to non-linearities and other practical influences in the motor's design and operating environment.

Despite these variations from the ideal models, the analytical approach detailed previously still offers substantial insight into the basic principles of motor operation. Understanding these foundational concepts is crucial, as it lays the groundwork for more sophisticated analyses. However, a more rigorous and detailed examination is necessary to accurately predict and replicate the actual behavior of motor waveforms under real-world conditions. This deeper level of analysis is essential for designing, testing, and optimizing motors to meet specific performance criteria and to ensure efficient and reliable operation in their intended applications.

In Fig. 2.12, two more representative and realistic waveform sets are depicted, illustrating the variations that can occur due to different motor designs and winding configurations. For motors with full pitch windings, such as those discussed in this chapter, the waveform of the back electromotive force (EMF) tends to assume a more trapezoidal shape. This characteristic shape, highlighted by the lighter curves in the figure, arises largely because of flux leakage that occurs as magnetic flux moves from one magnet to another within the motor structure. On the other hand, other motor designs aim for a flux linkage, back EMF, and torque that follow sinusoidal patterns, which are represented by the darker curves in the figure. These sinusoidal characteristics are typical of

Fig. 2.12 Wave forms that typically produce motors

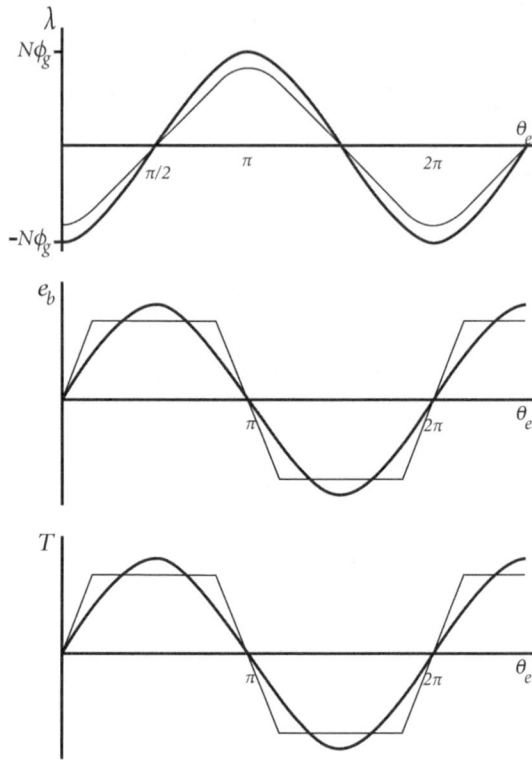

motors that employ winding configurations other than full pitch. The choice of winding and the specific magnetic flux management in these designs help in achieving a smoother, more continuous waveform that is ideal for applications requiring uniform motion and reduced torque ripple. These examples underscore the impact of motor design choices on the operational characteristics and efficiency of the machine, highlighting the importance of tailoring motor configurations to meet specific application needs.

2.10 Multiple Coils

In the typical design of a conventional motor, there often remains unused potential due to unutilized space that could otherwise accommodate more coils. This underutilization also includes the flux from three magnet poles that goes unused continuously [11]. To enhance motor performance, the design can be modified to include three additional full pitch coils, as illustrated in Figs. 2.13 and 2.14. This modification involves the introduction of two extra slots, thereby increasing the total number of slots to allow for the inclusion of three new coils. Consequently, each slot now houses two sides of a coil instead of one.

Fig. 2.13 Four full-pitch coils of a motor

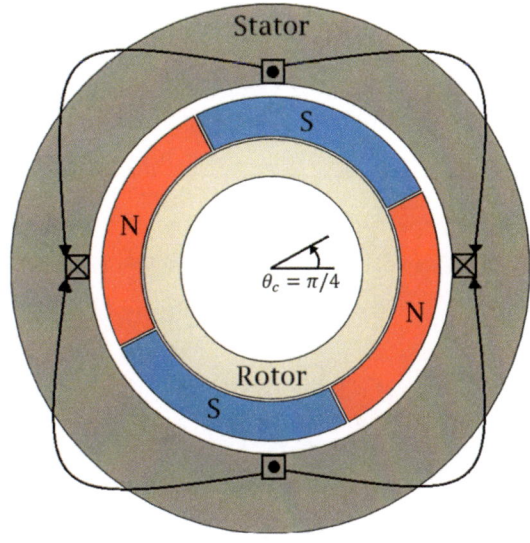

Fig. 2.14 Motor and its three phases

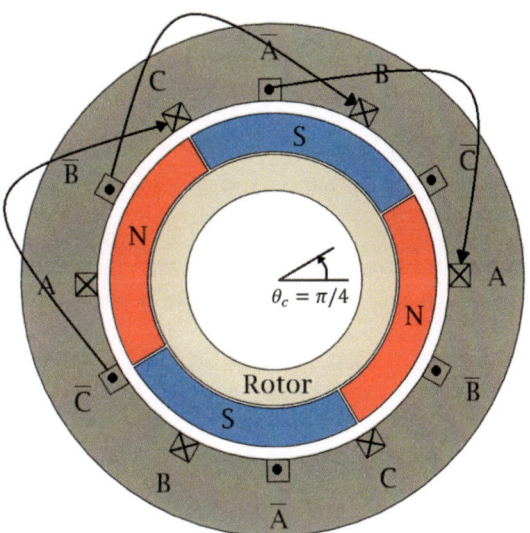

The arrangement of these coils is carefully considered to maximize flux linkage; starting with the first coil and moving counterclockwise around the stator, each successive coil is wound in the direction opposite to the previous one. This winding pattern ensures that the flux linkage for each coil mirrors that of its predecessor due to the alternating North and South magnet poles. The configuration offers flexibility in how the individual coils are connected, leading to various possibilities for grouping the coils into what is termed a phase winding or simply a phase.

Typically, all coils in a phase are connected in series, meaning the end of one coil is linked to the start of the next. This series connection allows the back EMFs from each coil to accumulate, contributing to a collective back EMF for the entire winding. Given that each coil produces a back EMF with a similar shape, the overall amplitude of the back EMF for the winding, as originally defined in Eq. 2.16, is now adjusted according to the new configuration as described in Eq. 2.22. This approach effectively enhances the motor's output by utilizing the magnetic flux more comprehensively and increasing the efficiency of the electromagnetic interactions within the motor.

$$E_b = 2N_m 2NB_g L_{st} R_{r0} \omega_m \tag{2.22}$$

Under such conditions de extend of the torque depends on N_m as it is shown in the following expression:

$$|T| = 2N_m 2NB_g L_{st} R_{r0} i \tag{2.23}$$

This formula establishes the relationship expressed as $T = kD^2 L$. In this equation, one instance of the diameter D is represented as the radius R_{r0}, and the other diameter is suggested by the number of magnet poles N_m. Additionally, the length L is directly represented as L_{st} in the equation, illustrating how these physical dimensions of the motor influence the torque calculation.

2.11   Effects of Multiple Phases

The motor discussed previously and depicted in Fig. 2.15 is classified as a single-phase motor. This type of motor is relatively uncommon in many applications due to its inability to generate torque consistently at all rotor positions [12]. Specifically, every 180°E, both the back electromotive force (EMF) and torque pass through zero, at which points the motor is incapable of producing torque. Furthermore, if the motor happens to stop at these zero-crossing points, it cannot restart on its own; the shaft must be manually rotated to a position where torque generation is possible. To overcome these limitations and facilitate the production of constant torque, brushless permanent magnet motors are typically equipped with more than one phase winding.

These additional phase windings are strategically oriented to ensure that their points of zero back EMF and torque are evenly distributed across an electrical period. Most commonly, these motors are designed with three phases, although configurations with one or two phases also exist. However, incorporating more than three phases is generally rare and only considered practical at very high-power levels. This is due to the increased number of power electronics required to drive additional phases, which complicates the system and elevates costs unless the application justifies the need for multiple banks of power electronics.

Fig. 2.15 Torque and back
EMF waveforms for a motor

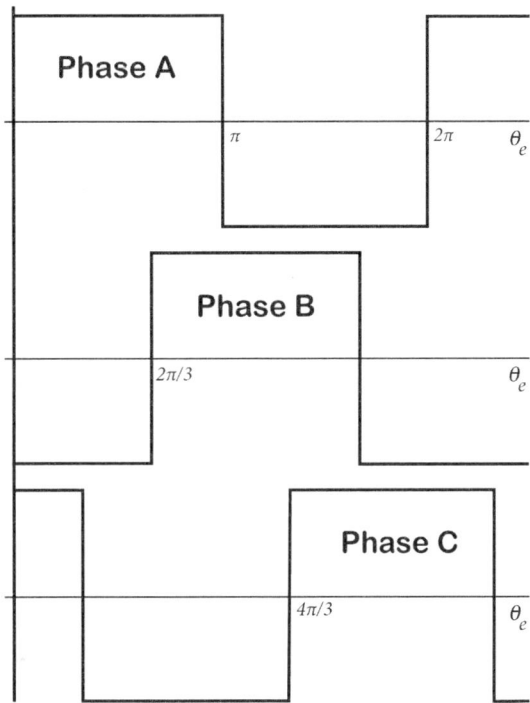

The motor configuration illustrated in Fig. 2.14 is designed to support the addition of
two more-phase windings, mirroring the construction of the initial phase winding with
each comprising four coils. This expansion results in a three-phase motor configuration,
as depicted in Fig. 2.15. In this figure, the phases are labeled A, B, and C. To maintain
clarity and avoid visual clutter in the diagram, Fig. 2.15 displays only one coil per phase.
The presence of the additional coils for each phase is implied through the phase labels
strategically placed near each slot. This method of depiction simplifies the understanding
of the motor's layout while clearly indicating the comprehensive multi-phase structure,
which significantly enhances the motor's operational capability and efficiency by ensuring
continuous torque production across various rotor positions.

The phase A winding depicted in Fig. 2.15 aligns with the configuration previously
detailed in Fig. 2.14, adhering to a systematic design approach across multiple phases.
With three phases incorporated into the motor, the back electromotive force (EMF) and
torque experience zero crossings that are equidistantly separated by one-third of a full
circle, equating to 120°E, or 60°M. This precise spacing dictates that the slots allocated
for phase B are positioned 60°M from those of phase A, and similarly, the slots for phase
C are rotated another 60°M from those of phase B.

This strategic placement effectively distributes slots at intervals of 30°M around the circumference of the stator, as illustrated in the figure. The ripple effect of this configuration is evident in the phase waveforms presented in Fig. 2.15, where the geometric and temporal alignment of the slots plays a critical role in shaping the motor's electromagnetic responses. The rotor remains constant across configurations, ensuring that any variations in flux linkage, back EMF, and torque across phases are solely attributable to the angular offset in slot placement. Consequently, the waveforms for phase B mirror those of phase A but are chronologically shifted by 120°E, reflecting the physical displacement between their respective slots. Similarly, the waveforms for phase C replicate those of phase A, yet they are displaced by a cumulative 240°E. This pattern of displacement ensures that each phase contributes uniformly to the motor's overall functionality while maintaining a cohesive and balanced operation.

This configuration not only distributes the mechanical components evenly but also ensures that the phase waveforms, shown in another figure (Fig. 2.16), are evenly offset to maintain continuous motor operation. As a result, each phase's flux linkage, back EMF, and torque maintain the same waveform shape, yet they are sequentially delayed in accordance with their physical displacement on the stator. Specifically, the waveform for phase B replicates that of phase A but is delayed by 120°E, and the waveform for phase C follows the same pattern, delayed by 240°E relative to phase A. This delay corresponds directly to their slot placement, ensuring that the motor provides balanced and uninterrupted torque production as the rotor turns.

Fig. 2.16 Fractional pitch coil

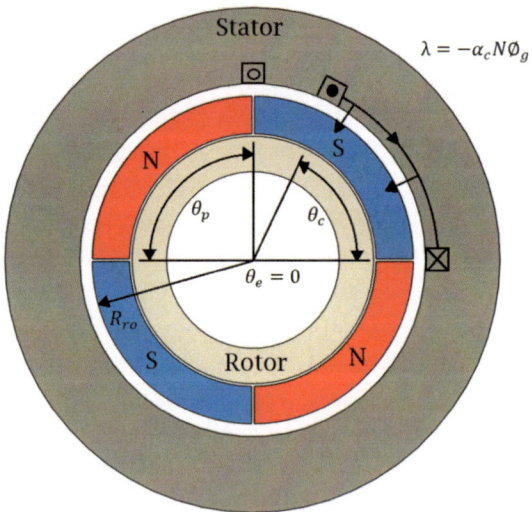

2.12   Different Designs

The motor previously discussed featured full pitch coils and magnet poles that extended over 180°E. Moreover, all four coils within each phase were aligned such that the back EMFs produced by each coil were in phase with one another. Consequently, when these coils were connected in series, the resultant back EMF preserved the waveform shape characteristic of a single coil but exhibited an amplitude that was four times greater. This arrangement simplifies the analysis of the motor's electromagnetic behavior due to the uniformity and predictability of its electrical outputs across the phases. However, it's important to note that such configurations are not commonly found in actual motor designs. Real-world motors often deviate from this ideal due to variations in coil pitch, magnetic pole coverage, and coil alignment within phases. These variations can significantly affect the performance and operational characteristics of the motor, making it crucial to explore how these differences influence motor behavior. In this section, we delve into how changes in these parameters impact the motor's functionality and efficiency.

When the coil pitch is less than 180°E, the resulting configuration is known as a fractional pitch winding. This type of winding alters how the back electromotive force (EMF) and torque are generated within the motor. To explore the effects of a fractional pitch winding, consider the motor depicted in Fig. 2.17. In this illustration, the rotor configuration remains the same, but the coil now spans a reduced angular coil pitch of $\theta_c = 180°E$, with the coil pitch factor defined as $\alpha_c = \theta_c/\theta_p$. At the starting position $\theta_c = 0$ shown in the figure, the flux linkage resulting from this fractional pitch configuration exhibits a smaller amplitude compared to that of a full pitch coil. This reduction in amplitude is directly proportional to the coil pitch factor α_c, illustrating how adjusting the coil pitch can influence the motor's electromagnetic output. Such changes impact not only the amplitude of the back EMF and torque but also their distribution and efficiency across the motor's operation.

As the rotor advances to $\theta_e = 60°E$ or $\pi/3$ radE, depicted in Fig. 2.18, the flux linkage experiences an increase from its minimum value to zero. This occurs because the flux traveling through the coil is balanced in both directions. Continuing the rotation, when the rotor aligns to $\theta_c = 120°E$ or $2\pi/3$ radE, as shown in Fig. 2.19, the flux linkage peaks since the coil is primarily facing the North magnet pole. This maximum level of flux linkage is sustained until the rotor reaches $\theta_c = 180°E$, at which point it begins to decline, crossing through zero and descending to the minimum once again.

Further rotation of the rotor reveals the complete cycle of the flux linkage waveform. In Fig. 2.20, this waveform, along with the resultant back EMF, is superimposed onto the waveforms from the ideal full pitch case for comparison. Notably, the back EMF in this fractional pitch setup includes segments where it remains at zero, corresponding to positions where the flux linkage remains constant. These zero segments in the back EMF translate directly to similar zero segments in the torque waveform, given that the shape of the back EMF and torque are congruent. This characteristic underscores how variations

Fig. 2.17 Rotor al 60° E

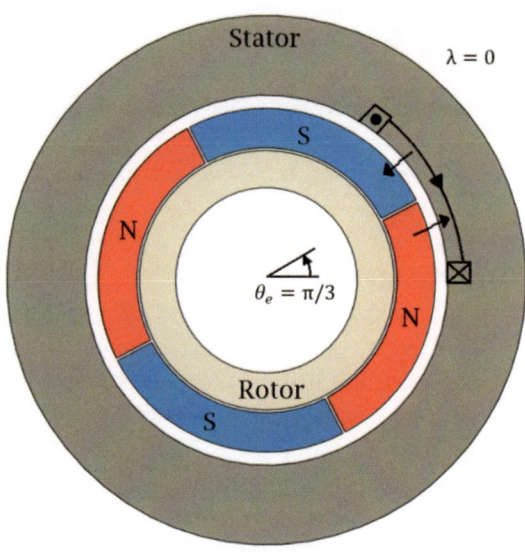

Fig. 2.18 Rotor al 120° E

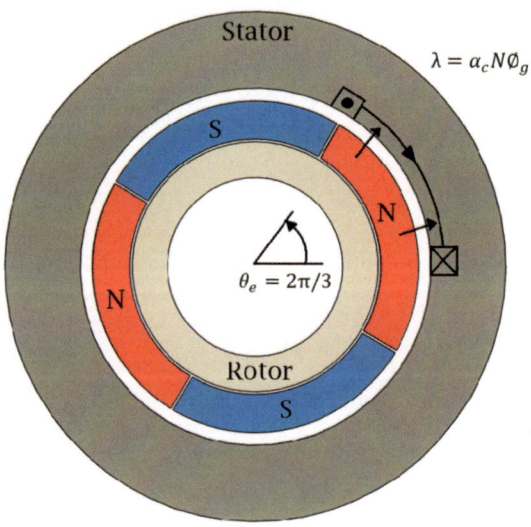

in coil pitch not only influence the magnitude but also the continuity of both back EMF and torque, affecting the motor's operational efficiency and smoothness.

As observed in the current motor design, three out of the four magnets are not actively engaged in torque production, which points to an inefficiency in the motor's design. To optimize the performance of the motor, additional coils should be incorporated to effectively utilize these inactive magnets. These added coils, if designed with the same fractional pitch as the existing coil, need to be strategically placed around the stator. This

Fig. 2.19 Back EMF and flux linkage in case of a pitch coil

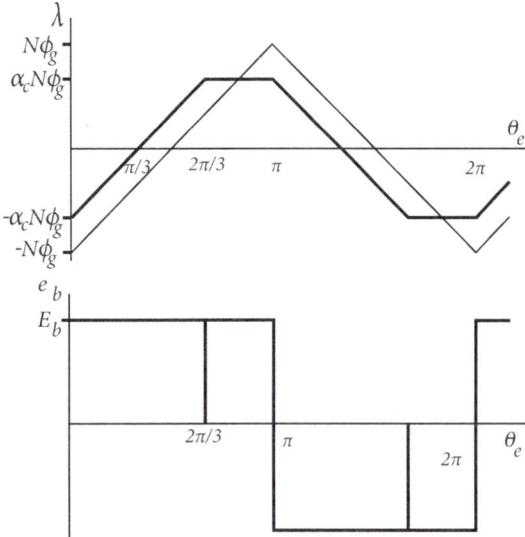

Fig. 2.20 Four fractional pitch coils for one phase

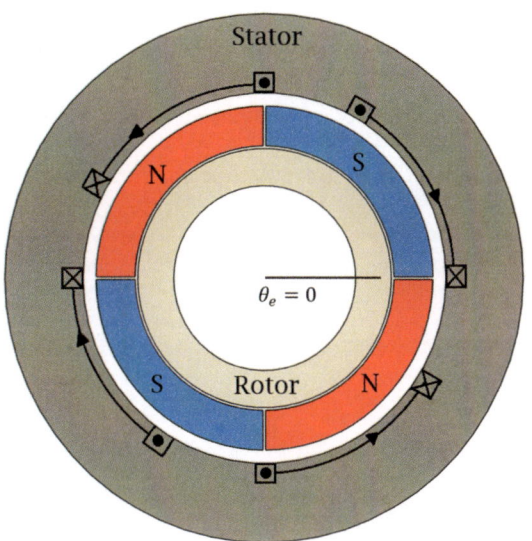

placement ensures that the back EMF generated by each new coil constructively adds to the overall net back EMF, enhancing the motor's performance.

Figure 2.20 illustrates a potential configuration for placing these additional coils. According to this arrangement, one side of each new coil aligns with a transition point between magnet poles, and every alternating coil is wound in the opposite direction to maintain a balanced and effective magnetic interaction. If these coils are then connected in series to form a single phase winding, the resultant back EMF will mirror the wave-form shown in Fig. 2.20 but will be amplified, with an amplitude that is N_m times greater, where N_m represents the number of magnets utilized. This approach not only increases the motor's torque output but also maximizes the use of all available magnets, leading to more efficient motor operation.

Magnets in practical motor designs rarely span the full pole pitch of 180°E because the flux at the transitions between North and South poles doesn't effectively contribute to torque production. Instead, this transition flux typically passes directly from one pole to another without interacting with the coils in the stator slots, leading to inefficient use of magnet material if full pitch magnets are employed. This inefficiency prompts the use of fractional pitch magnets to optimize material usage and performance.

To explore the implications of using fractional pitch magnets, consider the motor con-figuration illustrated in Fig. 2.21, which features a full pitch coil setup but includes gaps filled with nonmagnetic material between the magnet poles. Consequently, the angular magnet pitch (θ_m) is less than the angular pole pitch (θ_p), which alters the way flux dis-tribution and density are calculated. In this setup, the cross-sectional areas for the magnet and air gap are calculated as $R_{ro}\theta_m L_{st}$ rather than $R_{ro}\theta_p L_{st}$, reflecting the actual magnet coverage rather than the theoretical full pitch.

The expression for air gap flux density as given in Eq. 2.15 remains applicable but is now specific to the flux density over the surfaces where magnets are actually present. Under the assumptions of this scenario, no flux crosses the air gap in the regions corre-sponding to the gaps between the magnets. This specific distribution must be accounted for in calculations and motor design to ensure accurate predictions of motor performance

Fig. 2.21 Fractional pitch magnets

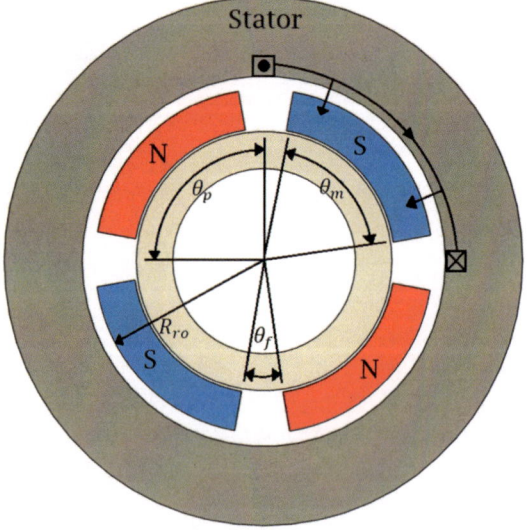

Fig. 2.22 Back EMF and flux
linkage in case of a pitch
magnet

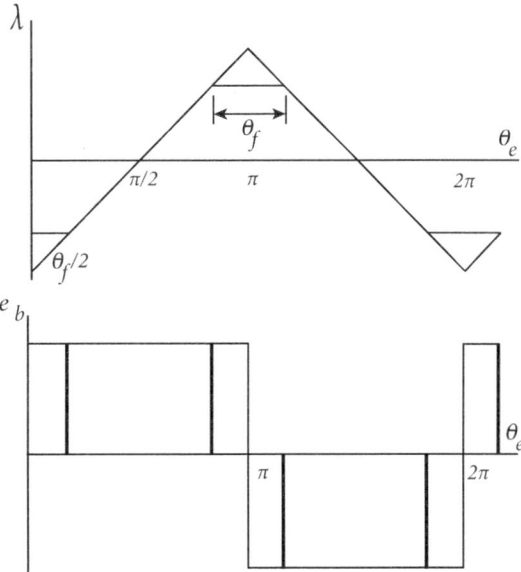

and efficiency. This approach minimizes wasted magnet material and enhances the overall
effectiveness of the motor.

The motor depicted in Fig. 2.15 is equipped with $N_s = 12$ slots, $N_m = 4$ magnet poles,
and $N_{ph} = 3$ phases, resulting in a configuration where $N_{ssp} = \frac{N_s}{N_m N_{ph}} = 1$, indicating one
slot per pole per phase. Although not all slots are visible in the figure, it is implied that
they are uniformly distributed around the stator at 30°M intervals, similar to the motor
shown in Fig. 2.21, which also has $N_{ssp} = 1$. In Fig. 2.21, despite the winding being of
fractional pitch, the slots are arranged just as in Fig. 2.22.

When N_{ssp} is an integer, the motor is classified as an integral slot motor. Conversely,
when N_{ssp} includes a fractional component, the motor is known as a fractional slot motor.
It's important to distinguish between motors with fractional pitch windings and fractional
slot motors. The former refers to the pitch of the windings within the slots, highlighting
how the coils are spatially configured relative to the magnetic poles. The latter concerns
the actual distribution of the slots around the motor's stator, which affects how the windings are laid out and interact with the magnetic field. Both characteristics have significant
implications for motor design and performance, influencing factors like torque production
and efficiency.

In an integral slot motor, a key characteristic is that the back electromotive forces
(EMFs) of all coils comprising a phase winding align in phase with one another. This
synchronization occurs even if the winding itself adopts a fractional pitch, as depicted
in Fig. 2.23. Such alignment facilitates a straightforward additive effect when coils are
connected in series, where the resultant back EMF essentially mirrors the waveform of a
single coil, merely scaled up in amplitude according to the number of coils.

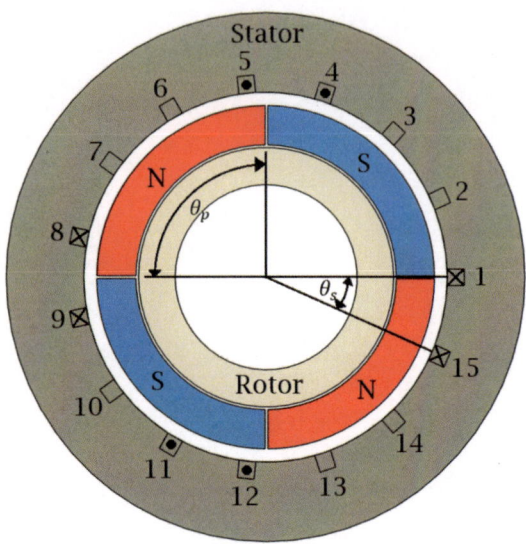

Fig. 2.23 The fractional visualization of a motor

Contrastingly, in a fractional slot motor, the back EMF generated by the coils of a phase winding does not uniformly align in phase. This misalignment is due to the staggered placement of coils relative to the magnet poles, a result of the non-integer slot per pole per phase ratio. As a consequence, when these coils are connected in series, the combined back EMF does not simply scale up in amplitude. Instead, it assumes a different shape and amplitude altogether. This divergence in waveform characteristics allows designers more flexibility in shaping the motor's back EMF according to specific performance criteria. The ability to manipulate the back EMF waveform in fractional slot motors is a distinct advantage, offering the potential to optimize motor characteristics such as torque smoothness, efficiency, and responsiveness to control inputs.

To grasp the distinctions between a fractional slot motor and an integral slot motor, let's examine the motor configuration presented in Fig. 2.1. This particular motor is equipped with $N_s = 15$ slots, $N_m = 4$ magnet poles, and $N_{ph} = 3$ phases. Calculating the number of slots per pole per phase (N_{spp}) for this setup results in a value of 1.25, indicating a fractional slot arrangement. This is derived from dividing the total number of slots by the product of the number of poles and phases, specifically $15/(4 \times 3)15/(4 \times 3)$.

Additionally, the angular slot pitch (θ_s) can be determined by the formula $360°M/N_s$, which calculates the angular distance between adjacent slots. For this motor, θ_s equates to 24°M or 48°E, showing that each slot is spaced at these angular intervals around the motor's stator. This setup, characterized by a non-integer N_{spp} and the specific angular slot pitch, typifies a fractional slot motor, which, unlike an integral slot motor, does not align the back EMFs of all coils in a phase. This fractional configuration allows for a more nuanced control over the electromagnetic properties of the motor, potentially enhancing performance but also complicating the waveform interactions of the back EMF and, consequently, the overall motor behavior. This understanding is crucial for designing and optimizing motors for specific applications where control precision and efficiency are paramount.

Due to the slot pitch not being an integral subdivision of the angular pole pitch (θ_p), utilizing full pitch coils in the described motor setup is impractical. The slot pitch of 48°E complicates how the motor should be wound, as it doesn't neatly align with the pole pitch of 180°E. If a coil spans three slots, the resulting angular coil pitch would be 3 · 48°E or 144°E, which is less than the angular pole pitch. Conversely, spanning four slots would create a coil pitch of 4 · 48°E or 192°E, exceeding the pole pitch. This discrepancy between the slot and pole pitches presents a challenge in determining the optimal coil span for maintaining effective electromagnetic interactions within the motor.

For illustrative purposes, a specific winding layout for phase A is shown in Fig. 2.24, deferring the detailed discussion of winding layout design to a later chapter. Given that the motor has 3 phases and 15 slots, with each coil effectively occupying one slot (since each coil side occupies half a slot), the calculation yields $N_{cph} = \frac{N_s}{N_{ph}} = \frac{15}{3} = 5$ coils per phase. Each illustrated coil spans three slots, giving a coil pitch or throw of three slots. This configuration allows for a coherent demonstration of how coils can be effectively arranged within the constraints of the motor's slot and phase structure, providing a foundational layout from which more complex or optimized winding configurations might be developed based on specific performance criteria or motor characteristics.

The analysis method previously outlined is applied in Fig. 2.25 to demonstrate the flux linkage and back electromotive force (EMF) for the coil labeled C_a. Given that the other coils in the setup span the same distance as coil C_a, it follows that their flux linkages

Fig. 2.24 Motor of 15 slots wit a phase A considering 4 poles

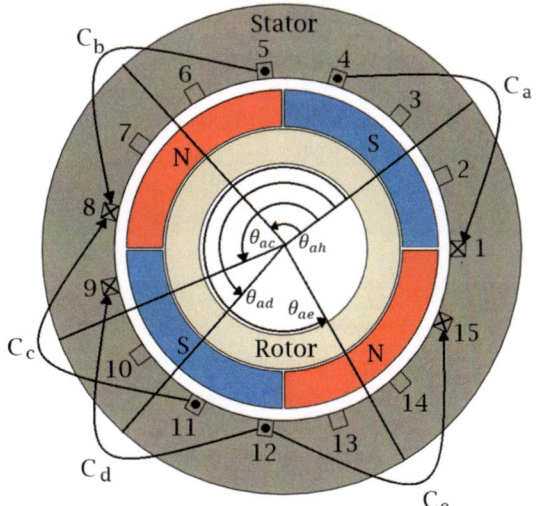

Fig. 2.25 Back EMF and flux linkage for the coil C_8

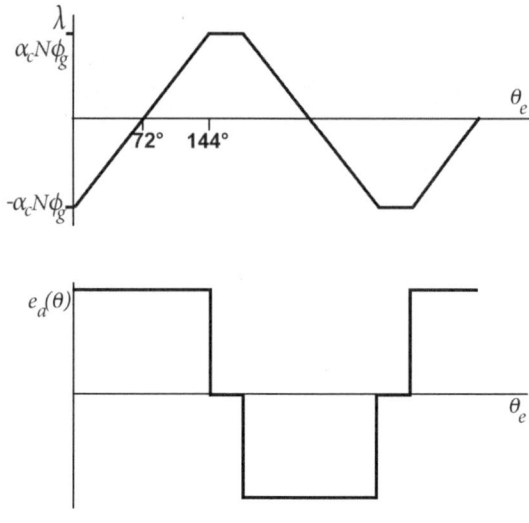

and back EMFs will exhibit the same shape and amplitude. However, these values will be phase-shifted relative to coil C_a due to their distinct angular positions within the motor's structure. This phase shift is calculated based on the angular displacement from the center of coil C_a, as exemplified in Fig. 2.24.

For instance, the center of coil C_b is offset by 4 slots from C_a, resulting in an angular displacement (θ_{ab}) of $4 \cdot 48°E = 192°E$. Similarly, for other coils, $\theta_{ac} = 7 \cdot 48°E = 336°E$, $\theta_{ad} = 8 \cdot 48°E = 384°E$, and $\theta_{ae} = 11 \cdot 48°E = 528°E$. It's also important to note that coils C_b and C_e are wound in the opposite direction to coil C_a. This opposite winding results in their back EMFs having opposite signs compared to that of coil C_a. This inversion of the back EMF sign for certain coils, coupled with the phase shifts due to their respective placements, highlights the complex interplay of physical and electrical properties within the motor, directly influencing its operational characteristics and performance.

Using the back EMF data from coil C_a as shown in Fig. 2.25, denoted as $e_a(\theta)$, we can deduce the back EMF relationships for the other coils in the phase based on their phase shifts and winding directions. For instance, coil C_b, which is wound in the opposite direction to C_a and is shifted by $192°$, has a back EMF represented as $-e_a(\theta - 192°)$. Similarly, coil C_c has a back EMF of $e_a(\theta - 336°)$, coil C_d follows with $e_a(\theta - 384°)$, and coil C_e, also wound oppositely to C_a, has $-e_a(\theta - 582°)$.

When these five coils are connected in series, the overall or net winding back EMF for the phase is the algebraic sum of the individual coil back EMFs. This configuration ensures that the EMFs contributed by each coil are properly aligned considering their phase shifts and direction of winding, effectively combining their electrical outputs. This summation of the back EMFs, which results in a more complex waveform, is precisely modeled in Eq. 2.24. The way these individual EMFs are added reflects the cumulative electromagnetic effect of the entire winding setup, illustrating how series connections of coils with varying orientations and phase positions can be optimized to achieve desired electrical characteristics in the motor.

$$e(\theta) = e_a(\theta) - e_a(\theta - 192°) + e_a(\theta - 336°) + e_a(\theta - 384°) - e_a(\theta - 582°)$$
$$= e_a(\theta) + e_a(\theta - 12°) + e_a(\theta + 24°) + e_a(\theta - 24°) - e_a(\theta + 12°) \qquad (2.24)$$

The equation provided utilizes the properties $-f(\theta) = f(\theta \pm 180°)$ and $f(\theta) = f(\theta \pm 360°)$ to streamline the initial expression. This simplification is pivotal as it clarifies that the net winding back electromotive force (EMF) comprises the sum of five phase A back EMFs, each respectively shifted by 0, $\pm 12°$E, and $\pm 24°$E. These shifts represent phase adjustments necessary to align the outputs of the individual coils when connected in series.

In Fig. 2.25, the back EMF for phase A coil C_a is detailed, and Fig. 2.26 subsequently illustrates the net winding back EMF resulting from combining these five phase-shifted EMFs. The first five waveforms displayed in the figure correspond to the individual back EMFs from the coils, and the final waveform represents the aggregated net back EMF. Interestingly, this combined waveform appears significantly more sinusoidal than the waveforms of the individual coils. This outcome suggests that when multiple coils are synchronized and their outputs combined, the overall waveform smoothens, reducing the angular disparities seen in single coil outputs.

This phenomenon of a more sinusoidal waveform in the final net back EMF, especially relevant in a practical motor setup like a four-pole, fifteen-slot motor, is often enhanced by the inherent leakage flux, which, although not accounted for in this particular theoretical analysis, plays a crucial role in real-world applications. Leakage flux tends to smooth out irregularities and discrepancies in magnetic flux distribution, contributing to a more uniform and sinusoidal back EMF across the motor's operational spectrum.

2.13   Coil Resistance

In all motor designs, multiple coils are interconnected to form what are known as phases, which are fundamental to motor functionality. Coils possess two key electrical properties: resistance and inductance [13]. Resistance, the simpler of the two to quantify, is an inherent property of all materials and acts as a measure of the material's opposition to the flow of electric current. Specifically, in conductive materials like copper, the resistance R can be precisely calculated in ohms (Ω) based on characteristics such as the material's resistivity, length, and cross-sectional area, as detailed by Eq. 2.25.

Inductance, on the other hand, presents a more complex property to describe, particularly because it often involves mutual inductance between coils, a common scenario in motor assemblies. Inductance is crucial in determining how a coil will react to changes

Fig. 2.26 Winding back EMF

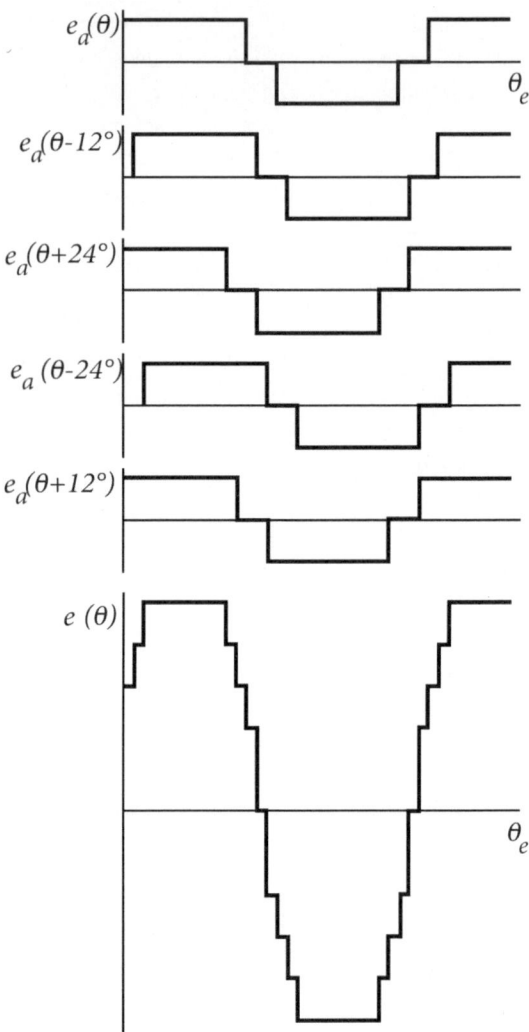

in current, affecting how energy is stored in magnetic fields during operation. The complexity of inductance becomes more pronounced with the inclusion of mutual inductance, where the magnetic field of one coil influences others, complicating calculations and impacting overall motor performance. Thus, while resistance provides a straightforward measurement of how easily current flows through a material, inductance involves dynamic interactions that are integral to understanding a motor's electromagnetic behavior.

$$R = \frac{\rho l}{A} \tag{2.25}$$

The resistance R of a material is calculated using the formula where ρ represents the resistivity of the material, measured in ohms-meter $(\Omega \cdot m)$, l is the length of the material through which current flows, and A is the cross-sectional area of the material. This formula effectively quantifies how much the material impedes the flow of electric current, with resistivity ρ playing a central role. In practical terms, resistivity is not a static value but varies with temperature—a critical consideration in electrical engineering. Typically, resistivity increases as temperature rises, often following an exponential trend for many materials.

However, for conductive metals such as copper and aluminum, commonly used in wiring, this relationship can be simplified and is generally approximated by a linear model, as depicted in Eq. 2.26. This linear approximation is particularly useful for practical calculations and engineering applications, allowing for straightforward adjustments based on changes in temperature. This relationship highlights the sensitivity of metal conductors to temperature changes, influencing their performance characteristics in electrical circuits and systems, including those found in motors and other electrical machinery.

$$\rho(T) = \rho(T_0)(1 + \alpha(T - T_0)) \tag{2.26}$$

In this equation, T represents the current temperature, T_0 is a reference or base temperature, and α is the thermal resistivity coefficient, which quantifies how much the resistivity of the material changes with temperature. For copper, which is widely used in electrical applications, these parameters are typically set with $T_0 = 20$ °C, the resistivity at this temperature $\rho(T_0)$ as $1.7241 \times 10^{-8} \Omega \cdot m$, and α at 4×10^{-3} per degree Celsius. This setup implies that as the temperature of copper increases, its resistivity also rises linearly according to this coefficient.

The implication of the coefficient α is significant: for copper, the resistance increases by approximately 4% for every 10 °C rise in temperature. This relationship means that if the temperature of a copper wire increases from 20 °C to 120 °C, the resistance of the wire would increase by 40% compared to its resistance at the base temperature. It's important to note, however, that this linear approximation starts to become less accurate as the material temperature exceeds 100 °C. Beyond this point, the actual resistivity tends to be higher than what the equation predicts, indicating that the linear model underestimates the increase in resistivity at higher temperatures. This aspect is crucial for applications involving significant temperature variations, as the actual performance may differ from the predicted performance due to these variations in resistivity.

The relationships described above underpin the typical construction of motor coils, which are generally made up of multiple turns of round, insulated wire. This configuration is depicted in Fig. 2.27. At the core of this setup is the bare conductor, characterized by a diameter d_{wv} and a cross-sectional area A_{wv}. Surrounding the conductor is the wire insulation, which plays a critical role in preventing electrical shorts between the turns of the coil. This insulation is commonly available in three thickness levels: single, double (or heavy), and triple, each providing varying degrees of protection and durability.

Fig. 2.27 Transversal section
of a wire

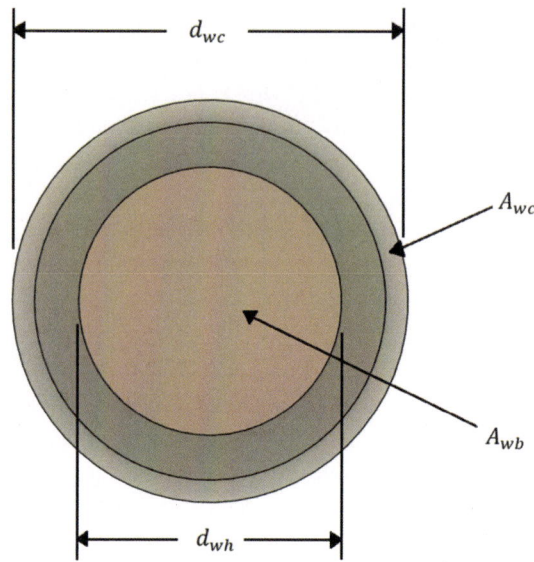

Over the insulation, there may be an optional bonding layer, typically made of an adhesive material. This layer's primary purpose is to bond the turns of wire together once the coil is formed, ensuring structural integrity and reducing the likelihood of movement that could lead to wear or electrical noise. The diameter and cross-sectional area of the wire, including this bonding layer, are denoted as d_{wc} and A_{wc}, respectively. This entire assembly ensures that the coils are robust enough to withstand the operational stresses within a motor, including thermal expansion and mechanical vibrations, while maintaining the electrical isolation necessary for safe and efficient operation.

Several standards are utilized to classify wire diameters, one of the most prevalent being the American Wire Gauge (AWG). In the AWG system, standard wire diameters are arranged in a geometric progression, which can be described by the formula $d = 7.3481^G$, where G represents the integer wire gauge number, and d_{wv} is the diameter of the bare wire measured in millimeters. This system establishes an inverse relationship between the wire gauge and its diameter: as the gauge number increases, the diameter of the wire decreases. This reduction in diameter as gauge increases also implies an increase in resistance per unit length of the wire, a crucial factor in electrical and electronic applications where resistance impacts the performance.

Further elaborating, the resistance increase associated with higher gauge numbers (smaller diameters) is due to the thinner cross-sectional area available for current to pass through, which naturally has higher resistive properties. The inverse of the relationship that defines wire diameter as a function of gauge is formalized in Eq. 2.27. This equation provides a mathematical basis for converting back from a measured diameter to the corresponding AWG gauge, facilitating the selection and specification of wire based on performance requirements and physical dimensions in various applications.

$$G = \frac{\log\left(\frac{d_{wv}}{8.2514}\right)}{\log(0.8990)} \tag{2.27}$$

The American Wire Gauge (AWG) system, structured around a geometric progression, enables wire gauges to be related to each other by specific ratios. This relationship is depicted in Fig. 2.28, which plots the resistance of wires relative to a wire of any given gauge G, extending to gauges $G + 1$, $G + 2$, etc. One of the key highlights on this curve is observed at the gauge $G + 3$. Here, a wire of gauge $G + 3$ exhibits twice the resistance of a wire of gauge G. Consequently, pairing two wires of gauge $G + 3$ in parallel results in the same resistance as a single wire of gauge G.

The chart also identifies significant points at gauges $G - 1$ and $G + 3$. At gauge $G + 3$, the resistance is approximately 26% greater than at gauge G. Therefore, a single increment in wire gauge leads to a 26% increase in resistance and, assuming constant current, a corresponding increase in I^2R losses. Conversely, at gauge $G + 1$, the resistance decreases to about 79% of that at gauge G. This reduction implies that by decreasing the wire gauge by one, the I^2R losses for a fixed current are reduced to 79% of what they would be at gauge G. These relationships offer vital insights for electrical applications, influencing decisions regarding wire sizing based on performance needs and efficiency considerations, especially where resistance and power losses are critical factors.

The current-carrying capacity of a wire is fundamentally influenced by its cross-sectional area and the thermal conditions surrounding it. Heat generated within a resistor, described as I^2R loss per unit volume, can be mathematically expressed as ρJ^2, where J is the current density flowing through the material. Practical experience suggests that the maximum allowable current density for wires typically ranges from 1 to 10 Arms/mm^2. This range is illustrated in Fig. 2.29, which plots allowable RMS wire current against wire gauge.

Fig. 2.28 Wire gage and its relative resistance

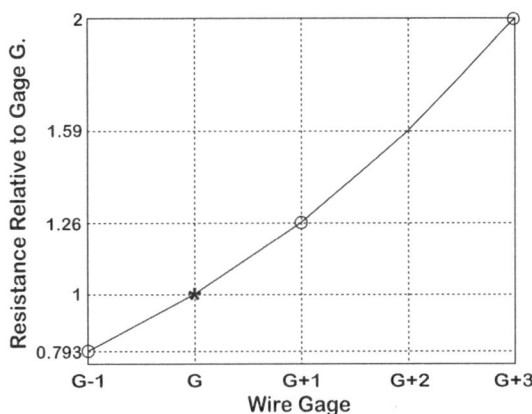

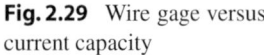

Fig. 2.29 Wire gage versus
current capacity

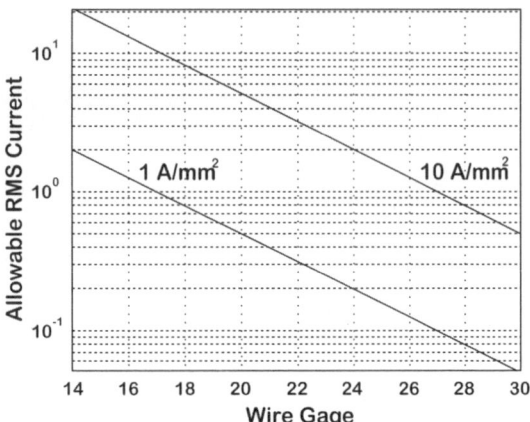

In environments where the wire is enclosed within confined spaces lacking adequate thermal conductivity, the minimum threshold of 1 Arms/mm^2 might pose a risk of overheating, suggesting that a lower limit may sometimes be necessary. Conversely, in scenarios where the wire benefits from active cooling, a maximum density of 10 Arms/mm^2 could be considered overly cautious, potentially limiting the efficiency of the system. An interesting application of these principles can be seen in household wiring, particularly with 14 gauge wire which is rated for 15 Arms. This rating translates to a current density of about 7.2 Arms/mm^2, situating it within the established range but highlighting the balance between safety, functionality, and thermal management in electrical wiring design.

In motor design, calculating the resistance of coils involves a straightforward application of the formula $d = 7.3481^G$ through Eq. 2.27 to determine the diameter of the bare wire based on its gauge. However, when accounting for the insulation and any optional bonding material, which add to the overall diameter, the computation becomes more complex. The thicknesses of the insulation and bonding material do not adhere to the geometric progression used for bare wire diameters. Thus, formulas like $d = 7.3481^G$ cannot be directly applied to determine the covered wire diameter (d_{wv}).

The variations in insulation (single, double, triple) and bonding material thicknesses are not standardized in relation to wire gauge and do not follow a predictable mathematical function. Therefore, these additional dimensions are best ascertained through specific manufacturer's wire tables or data sheets. These tables provide precise measurements for the total diameter of the wire including these layers, which is crucial for accurate design calculations in motor manufacturing. Consulting such detailed sources ensures that all factors affecting wire resistance, including non-conductive layers, are accurately incorporated into the design process.

2.14   Coil Inductance

In brushless permanent magnet motors, inductance is not typically considered a critical parameter. The primary role of inductance in these motors is to define the time constant of the windings, which influences how quickly winding currents can change [14]. However, modern power electronic control techniques, which often employ principles of current control, are generally less sensitive to the precise value of inductance. Therefore, in many cases, inductance calculations are simplified to provide analytical estimates that are adequate for most practical purposes.

If a more precise understanding of inductance is necessary, especially in scenarios where the coil configuration or motor design is complex, three-dimensional finite element analysis (FEA) is employed. This method offers a detailed assessment of the magnetic fields and their interactions within the motor's structure. When a coil is embedded within stator slots, its inductance significantly differs from what it would be if the coil were simply surrounded by air. This is due to the different magnetic permeability of the stator material compared to air, which can greatly influence the magnetic field distribution and thus the inductance. The behavior of some the elements are shown in Fig. 2.29.

Inductance can also be calculated by analyzing the energy relationships in the coil, specifically by equating the energy stored in the magnetic field (co-energy) with the energy stored in the inductance. This approach aligns with solving Eq. 2.28, where inductance values are derived based on the field dynamics created by the coil currents. Such calculations are crucial when designing motors that need to meet specific dynamic response criteria or when integrating the motor with sophisticated control systems.

$$\frac{1}{2}Li^2 = \int \frac{1}{2}\mu H^2 dV \qquad (2.28)$$

where L is the inductance.

In the context of motor design, understanding both self-inductance and mutual inductance is crucial, particularly when assessing how coils interact within a motor. Self-inductance concerns the magnetic flux linkage that a coil generates due to its own current, whereas mutual inductance involves the influence of one coil's magnetic field on another. Here, we focus on mutual inductance between coils within the same phase, although it's important to note that mutual inductance also exists between coils of different phases. However, this inter-phase mutual inductance is generally small compared to self-inductance—typically around 10%—and requires more complex calculations, hence it is not considered in this discussion for simplicity and practicality.

When coils are inserted into slots of a motor's stator, the overall inductance of each coil is influenced by three distinct components corresponding to the key areas where significant magnetic fields are generated by the coil currents. These areas include the air gap, the slots, and the end turns of the coils. Each of these regions contributes differently to the total inductance based on their magnetic properties and the spatial configuration of the coils. The ferromagnetic parts of the motor, such as the stator core, generally do not add to the inductance when their relative permeability is high, because they support the flux generated by the coils without generating significant additional linkage.

Additionally, it is crucial to recognize that the flux produced by permanent magnets in the motor does not affect the inductance measurements, as inductance specifically quantifies the linkage of magnetic flux resulting solely from the coil's current. From the perspective of the coils, the magnets can be seen as blocks of material with a relative permeability ($\mu_0 \mu_R$), which does not actively contribute to the generation of inductance. This distinction is vital for correctly modeling and understanding the electromagnetic behavior of brushless motors, where the precise calculation of inductance is essential for performance tuning and ensuring efficient operation.

References

1. D. C. Hanselman, *Brushless Permanent Magnet Motor Design*. The Writers' Collective (2003)
2. R. W. Postma, Force and torque margins for complex mechanical systems. In Proceedings of the 37th Aerospace Mechanisms Symposium, Johnson Space Flight Center (2004, May), pp. 107–118
3. G.H. Jang, J.W. Yoon, N.Y. Park, S.M. Jang, Torque and unbalanced magnetic force in a rotational unsymmetric brushless DC motors. IEEE Trans. Magn. **32**(5), 5157–5159 (1996)
4. E. Urata, On the torque generated in a servo valve torque motor using permanent magnets. Proc. Inst. Mech. Eng. C J. Mech. Eng. Sci. **221**(5), 519–525 (2007)
5. W. Bauer, W. Amrhein, Electrical design considerations for a bearingless axial-force/torque motor. IEEE Trans. Ind. Appl. **50**(4), 2512–2522 (2014)
6. J. W. Ahn, Switched reluctance motor. Torque Contr. 201–252 (2011)
7. N. Hemati, M.C. Leu, A complete model characterization of brushless DC motors. IEEE Trans. Ind. Appl. **28**(1), 172–180 (1992)
8. K. Nakamura, K. Saito, O. Ichinokura, Dynamic analysis of interior permanent magnet motor based on a magnetic circuit model. IEEE Trans. Magn. **39**(5), 3250–3252 (2003)
9. H.P. Chi, R.L. Lin, J.F. Chen, Simplified flux-linkage model for switched-reluctance motors. IEE Proc. Electric Power Appl. **152**(3), 577–583 (2005)
10. S.B. Ozturk, W.C. Alexander, H.A. Toliyat, Direct torque control of four-switch brushless DC motor with non-sinusoidal back EMF. IEEE Trans. Power Electron. **25**(2), 263–271 (2009)
11. M. Kaufhold, H. Aninger, M. Berth, J. Speck, M. Eberhardt, Electrical stress and failure mechanism of the winding insulation in PWM-inverter-fed low-voltage induction motors. IEEE Trans. Industr. Electron. **47**(2), 396–402 (2000)
12. S. Cataldi, A.T. Stanley, M.C. Miniaci, D. Sulzer, Interpreting the role of the striatum during multiple phases of motor learning. FEBS J. **289**(8), 2263–2281 (2022)
13. R. Oboe, F. Marcassa, G. Maiocchi, Hard disk drive with voltage-driven voice coil motor and model-based control. IEEE Trans. Magn. **41**(2), 784–790 (2005)
14. T.J.E. Miller, M.I. McGilp, D.A. Staton, J.J. Bremner, Calculation of inductance in permanent-magnet DC motors. IEE Proc. Electr. Power Appl. **146**(2), 129–137 (1999)

Pole and Slot Configurations for Efficient Brushless Permanent Magnet Motors

3

3.1 Windings

Brushless permanent magnet motors are versatile in design, allowing for an even number of magnet poles (N_m) and any number of stator slots (N_s). However, despite the vast possibilities this flexibility affords, only a select few combinations of magnet pole and slot counts truly optimize the use of the stator slots and enhance efficient torque production. This optimization is crucial for achieving high performance in motor operations, as it directly influences the motor's efficiency and effectiveness in energy conversion [1]. The current chapter is dedicated to unraveling the principles necessary to pinpoint these optimal pole and slot count combinations specifically for three-phase motors. By exploring these configurations, the chapter aims to guide designers in selecting the most effective designs for specific applications. Furthermore, it introduces a systematic procedure for designing the winding layout tailored to any of these identified optimal combinations. This approach not only simplifies the design process but also ensures that each motor configuration is ideally suited to meet its intended performance criteria, maximizing both the physical and electrical integration of the components within the motor structure.

Given the infinite potential combinations for pole and slot counts as well as winding layouts in brushless permanent magnet motors, it's essential to apply certain assumptions to narrow down the choices to those that yield efficient and practical designs. The assumptions outlined here focus on creating desirable winding configurations under specific constraints:

(a) The motor configuration is based on a three-phase system. This assumption simplifies the initial design framework, though the principles discussed can be straightforwardly adapted to motors with different phase counts.

© The Author(s), under exclusive license to Springer Nature Switzerland AG 2025

E. Cuevas et al., *DC Motors*, Synthesis Lectures on Engineering, Science, and Technology, https://doi.org/10.1007/978-3-031-64354-5_3

(b) All slots in the motor are utilized, which implies that the number of slots (N_s) is a multiple of the number of phases (N_{ph}). For three-phase motors, this means that N_s is always a multiple of three, ensuring that each phase can be evenly distributed across the stator.

(c) Each slot contains two coil sides, defining the motor as having a double-layer winding. This setup enhances the motor's electromagnetic interactions and helps in achieving a more uniform magnetic field distribution.

(d) Only balanced windings are considered, where the back EMF for phases B and C is precisely 120°E offset from the back EMF of phase A. This ensures that the motor operates smoothly and efficiently, with uniform torque output and reduced vibrations.

(e) The number of slots per pole per phase ($N_{spp} = N_s/N_m/N_{ph}$) is restricted to two or less. This constraint simplifies the winding layout and is typical for most motors. If N_{spp} exceeds two, it introduces additional complexity that rarely enhances motor performance and may complicate the winding process, especially when using a stator lamination designed for a different motor configuration.

(f) All coils are identical in terms of the number of turns and the span of slots they cover, leading to uniformity in coil size, resistance, and inductance across the motor. This uniformity ensures that each phase contributes equally to the motor's overall function, facilitating easier manufacturing and consistent performance.

Adhering to these assumptions not only streamlines the design process but also ensures that the resulting motors are capable of high performance and are easier to manufacture. While it is possible to design motors that do not conform to one or more of these criteria, such deviations often lead to increased complexity in winding or diminished motor performance.

3.2   Coil Span

As discussed in Chapter 4, the coil span or coil pitch refers to the circumferential extent of a coil within a motor. This can be measured in either mechanical or electrical terms, but in the context of slotted motors, describing coil span in terms of slots is particularly practical [2]. For instance, if a coil stretches from slot k to slot $k + 2$, the coil span is defined as 2 slots. Typically, the coil span should approximate but not exceed 180°E to optimize performance. This positioning maximizes the magnetic flux linkage to the coil, thereby enhancing the back electromotive force (EMF) induced in the coil, which is critical for efficient motor operation.

However, there are exceptions to this general guideline, particularly when the slot pitch—the angular distance between adjacent slots—surpasses 180°E. This scenario is more common in motors with an outer rotor configuration where the number of slots N_s is fewer than the number of magnet poles N_m. Under such circumstances, if the slot pitch exceeds 180°E, it becomes necessary to minimize the coil pitch to just one slot to maintain functional integrity and performance. This adjustment ensures that despite the larger slot pitch, the coil still effectively interacts with the magnetic field, although the flux linkage and induced back EMF may be somewhat less optimized compared to configurations adhering to the ideal coil span of close to 180°E.

The number of slots per magnetic pole, as specified in Eq. 3.1, can be used to determine the nominal coil span, as outlined previously.

$$N_{sm} = \frac{N_s}{N_m} \tag{3.1}$$

The value of N_{sm} determines the number of slots per 180°E. Consequently, the nominal coil span in slots is given by the integer portion of Eq. 3.1, as stated in Eq. 3.2.

$$S^* = \max\left(\text{integer}\left(\frac{N_s}{N_n}\right), 1\right) \tag{3.2}$$

The function max(·) returns the higher of its two inputs, while the function integer(·) extracts the integer component of its argument. The max(·) function is incorporated in Eq. 3.2 to guarantee that the span is no less than one slot when N_s is less than N_m.

Sometimes, the actual span of the winding varies from the specified span outlined in Eq. 3.2. When this occurs, the most frequently selected span is equal to $S^* - 1$. Reducing the span diminishes the length of the end turns and modifies the amplitude and harmonic content of the flux linkage and subsequent back EMF. In this case, the winding is referred to as being short-pitched or chorded.

3.3   Optimal Pole and Slot Configurations

In motor design, particularly for three-phase motors, not every combination of magnet poles and stator slots will align with specific winding assumptions. To satisfy these assumptions, the number of slots, for instance, must be a multiple of three [3]. This requirement ensures that each slot can accommodate two coil sides. Before delving into the complexities of winding layout, it's crucial to identify which combinations of magnet poles and slot counts result in valid windings that adhere to these criteria.

For a three-phase motor to function effectively, each phase winding must not only produce a back electromotive force (EMF) of the same amplitude and shape but also achieve a critical phase shift of 120°E relative to the other two phases. This specification ensures that the motor's winding is balanced, which is vital for the smooth and efficient operation of the motor. Achieving identical amplitude and shape in the phase back EMFs requires that the coils in each phase have the same number of turns, maintain the same coil span, and are uniformly distributed around the stator.

These winding specifications naturally lead to valid pole and slot counts being determined by their capacity to facilitate the necessary 120°E phase offset among the three phase windings. Thus, the interplay between the number of poles and slots is not just a matter of mechanical fitting but a deliberate configuration to meet electromagnetic performance standards. By ensuring these parameters are met, the resulting motor design is optimized for performance, highlighting the importance of precise engineering in the initial stages of motor development.

Referring to Fig. 3.1, let's consider the layout for a three-phase motor where specific slots are utilized for coil placement. If the first coil of phase A occupies slots 0 and S, with S representing the chosen coil span, then the first coil of phase B needs to be positioned in a way that ensures its placement adheres to a 120°E separation from phase A. Specifically, phase B's first coil would use slots k and $k + S$, where k is determined to ensure that slots 0 and k are separated by exactly 120°E. This separation could potentially include an additional 120°E+q 360°E, where q is any integer, to accommodate the circular nature of the slot arrangement. This ensures that the principal angle between slots 0 and k aligns with the required 120°E phase shift.

If such a slot k cannot be identified, it implies that the selected combination of pole and slot counts does not permit a balanced winding structure. However, once an appropriate slot k. is found, it establishes a phase offset for phase B, where every coil is shifted by k slots compared to the corresponding coil in phase A. This phase offset, denoted as $K_0 = k$, is critical as it guarantees that the back EMFs of each coil in phase B are phase-shifted by 120°E relative to those in phase A. This strategic arrangement ensures the necessary phase alignment and contributes to the overall balanced operation of the motor, where each phase complements the others to produce a smooth and efficient rotational output.

Fig. 3.1 Fifteen slots for a stator

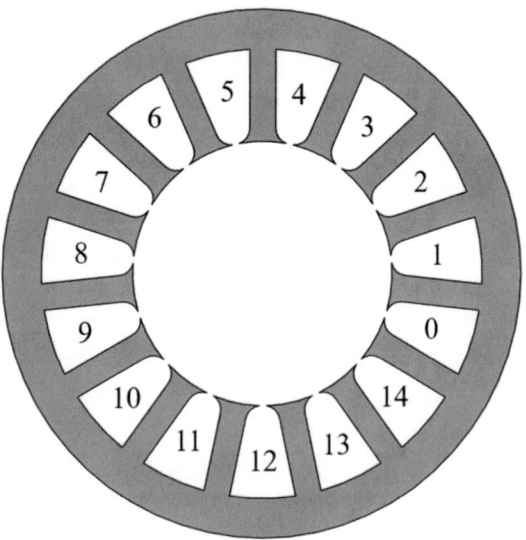

In the design of three-phase motors, achieving a balanced winding is crucial for optimal performance. This balance is accomplished by carefully determining the phase offsets between the coils of each phase. Specifically, if a phase offset of K_0 slots results in a 120°E offset between phases A and B, then similarly shifting the coils in phase C by K_0 slots from those of phase B ensures another 120°E offset. This sequential shifting effectively creates a balanced winding configuration where each phase complements the others, ensuring uniform torque and minimizing vibrations.

Mathematically, this phase offset K_0 can be precisely calculated by first identifying the angular position of each slot relative to a reference point, typically slot 0. The angular slot pitch, denoted as θ_s, is defined as $360°M/N_s$, where N_s is the total number of slots. Based on this definition, the angle of any slot k relative to slot 0 can be calculated using Eq. 3.3:

$$\theta_{sl} = k\frac{N_m}{2}\frac{360}{N_s} = k\frac{N_m}{N_s}180°\text{E}. \tag{3.3}$$

where $k = 1, \ldots, N_s - 1$. This equation provides a clear and systematic approach to determining the angular position of each slot, which is essential for setting the correct phase offsets. By applying this methodology, each phase's coils can be aligned to ensure that the requisite 120°E phase shifts are maintained throughout the motor, thereby achieving a balanced and efficient winding configuration. This balance is integral to the motor's functionality, reducing inefficiencies and enhancing overall system reliability.

To accurately determine the principal angle associated with each slot in a motor's configuration, the remainder function, often denoted as rem(x, y), becomes a critical tool. This function returns the remainder of the division of x by y, essentially calculating x modulo y. This method is especially useful when aligning the angles within a circular framework, such as the arrangement of slots in a motor, where the total angle for a complete revolution is 360°. By applying the remainder function, any slot angle that exceeds 360°—due to multiple rotations or high slot numbers—is effectively normalized back within the 0°–360° range. This calculation is formalized in Eq. 3.4:

$$\theta_{sl} = \text{rem}\left(k\frac{N_m}{N_s}180°\text{E}, 360°\text{E}\right). \tag{3.4}$$

Assuming that

$$\text{rem}\left(\frac{3N_m}{2N_s}K_0, 3\right) = 1 \tag{3.5}$$

By using this approach, each slot's angle can be accurately mapped onto the 360° cycle, aiding in the precise determination of slot positioning and phase offset adjustments in motor design. This mathematical treatment is essential for maintaining the integrity of angular measurements and ensuring that all calculations reflect the actual physical layout of the motor's stator slots.

In the context of motor design, particularly when aligning the phase windings for three-phase motors, determining the correct phase offset K_0 is crucial. K_0 is identified as the value of k (the slot number) for which the calculated angle from Eq. 3.5 equals 120 ° E. This equation essentially defines the angular displacement needed to achieve the necessary phase shift between each set of windings. It is entirely possible, given the number of slots and the configuration of the motor, that multiple values of k could satisfy this condition. When multiple solutions exist, any of them would technically work to establish the phase shift, but for simplicity and to maintain a compact and efficient design, the smallest value of k is usually selected.

To simplify the search for K_0, the equation can be adjusted by normalizing the angles involved. By dividing the terms in Eq. 3.5 by 120°E, we reformulate the problem to find the smallest integer k that satisfies this normalized condition. This approach streamlines the calculation process and ensures clarity in determining the most practical slot number to use as the phase offset, which is critical for achieving balanced and efficient motor operation. This method not only aids in the precise alignment of the windings but also optimizes the electrical performance of the motor by ensuring that the electromagnetic forces are evenly distributed across the phases.

Due to the characteristics of the remainder function, which calculates the remainder of division, formulating a closed-form solution for determining the phase offset K_0 in motor windings is not feasible. The remainder function does not lend itself to straightforward algebraic manipulation that would allow for such a solution. Nevertheless, if a balanced winding configuration is achievable, discovering this alignment typically involves a simple iterative process. By incrementally adjusting the values and checking the resulting phase shifts, one can systematically identify the correct phase offset.

Alternatively, a more direct approach can be adopted that bypasses the need for the remainder function by framing the problem differently. Instead of relying on the remainder function to normalize the angles, the equation can be adjusted to directly equate the calculated slot angle from Eq. 3.2 to 120°E+q 360°E where q is any integer. This formulation acknowledges the cyclic nature of angular measurements in a circle. Simplifying from this setup can lead to a more accessible expression for K_0, which is articulated in Eq. 3.6. This alternative method provides a clear pathway to calculating the phase offset by direct substitution and algebraic rearrangement, making it easier to apply in practical scenarios without iterative checks. This approach not only simplifies the calculation but ensures that the windings can be aligned precisely to maintain a balanced and efficient motor operation.

$$K_0 = \frac{2N_s}{3N_m}(1 + 3q) \tag{3.6}$$

In determining the validity of a phase offset K_0 for motor winding configurations, it is essential to apply Eq. 3.5 effectively. K_0 is deemed valid if, when evaluating Eq. 3.4 with an integer value for q within the range of 1 to $(N_m/2) - 1$, the outcome is an integer.

This criterion ensures that the calculated phase offset aligns correctly with the physical slot structure of the motor.

For illustrative purposes, let's consider a motor with four poles and twelve slots. By iterating Eq. 3.4 (which might be referenced as Eq. 3.6 in detailed discussions), it becomes evident that $K_0 = 2$. Consequently, if phase A's first coil is placed in slot 0, then phase B's first coil should start in slot 2, and phase C's first coil in slot 4. This placement pattern perfectly corresponds with the coil arrangement depicted in Fig. 2.12, validating the calculation.

Extending this approach to a slightly more complex scenario, such as a motor with four poles and fifteen slots, iterating the same equation identifies $K_0 = 10$. Thus, if phase A's first coil is in slot 0, then phase B's should begin in slot 10, and phase C's in slot 20. However, since there are only 15 slots, slot 20 must be recalculated within the modular constraint of 15 slots, yielding $\text{rem}(20, 15) = 5$. Therefore, phase C actually starts in slot 5. This method of recalculating slot numbers ensures that the winding layout remains consistent and balanced even when the initial calculation exceeds the total number of available slots, demonstrating the practical application of modular arithmetic in motor design.

3.4 Motor Winding Configuration

The process of determining the optimal placement of windings in motors with N_m magnet poles and N_s slots appears to be either straightforward or closely guarded within the industry, as evidenced by the scarcity of detailed published works on the subject [4]. In the literature that does exist, winding configurations are often described using generic terms such as lap, wave, concentric, and sinusoidally distributed, but these descriptions typically lack comprehensive details or clear methodologies for achieving valid winding arrangements. Further complicating the discourse are technical terms like distribution factor, pitch factor, and winding factor, which are used to discuss how different winding layouts affect the flux linkage and the shape of the resulting back EMF.

Historically, these terms and concepts were more relevant when motor design was primarily a manual process and were particularly applied to types of motors other than brushless permanent magnet motors. In the modern era, with the advent of sophisticated computational tools, many of these traditional terms and concepts have become less critical. They are either calculated differently or are no longer applicable to contemporary motor designs, particularly in the realm of brushless permanent magnet motors. This shift underscores a broader trend in engineering where digital tools and computational models streamline processes and render some older methodologies obsolete, allowing for more direct and efficient design practices.

The winding layout described here is designed as a double layer lap winding, which is commonly found in virtually all brushless permanent magnet motors. This particular configuration is not only manufacturable but also optimized to enhance motor performance. While it is possible to alter the presented winding layout to create a wave winding, potentially with a single layer, such modifications generally do not lead to performance improvements. According to the *BLv* (Back EMF vs. velocity) and *BLi* (Back EMF vs. current) laws, the distribution of coil end turns—the sections of the coil that extend from one slot to another—has no impact on back EMF or torque generation. Instead, these characteristics are primarily influenced by how the coils are placed within the slots.

The role of the end turns is fundamentally to facilitate the transfer of current from one slot to the next without contributing directly to the motor's electromagnetic forces. As such, the *BLv* and *BLi* laws do not address end turns because they do not affect the operational outputs of back EMF or torque. However, the layout of end turns does affect other important aspects such as coil resistance and inductance, and it plays a significant role in the manufacturability of the motor. Efficient design and arrangement of end turns can therefore enhance the overall practicality and durability of the motor, despite their non-involvement in generating electromagnetic forces.

The primary objective in designing a motor winding layout is to strategically place coils so that the relative angular midpoints of the coils are optimally positioned, aiming for 0°E and 180°E separations. This configuration maximizes the efficiency of electromagnetic interactions within the motor. Coils that are positioned close to 0°E are wound in one direction, while those near 180°E are wound in the reverse or opposite direction. This opposite winding is necessary because the direction of magnet flux reverses at 180°E.

For illustrative purposes, let's examine an integral slot pitch motor with four poles and twelve slots as shown in Fig. 3.2. It's important to note that the slot numbering in this figure starts at one, differing from the zero-based numbering shown in Fig. 3.1, but the principle remains the same. In this example, coils with midpoints at angles θ_1 and θ_3 are aligned at what is designated as 0°E and are wound in one direction. Conversely, coils positioned at midpoints θ_2 and θ_4 are 180°E apart from θ_1 and θ_3, respectively, and are thus wound in the opposite direction to match the reversed magnetic flux.

To clearly indicate the direction of coil winding, terms like "In" and "Out" are employed in the diagrams. "In" refers to a coil side that enters a slot, whereas "Out" signifies a coil side that exits a slot. This terminology helps clarify the winding direction and ensures that the coils are correctly configured to align with the motor's magnetic dynamics, enhancing both the motor's function and its efficiency.

In fractional slot motors, achieving perfect alignment of all coils at exactly 0°E or 180°E separation is impractical due to the non-integral relationship between the number of poles and slots. This limitation necessitates selecting coil placements that approximate these ideal separations as closely as possible. To address this, coils that are nearest to the 180°E mark are wound in the reverse or opposite direction. This method of winding effectively adjusts the coil's angular position, shifting it back towards 0°E, aligning more closely with the desired electromagnetic interactions within the motor.

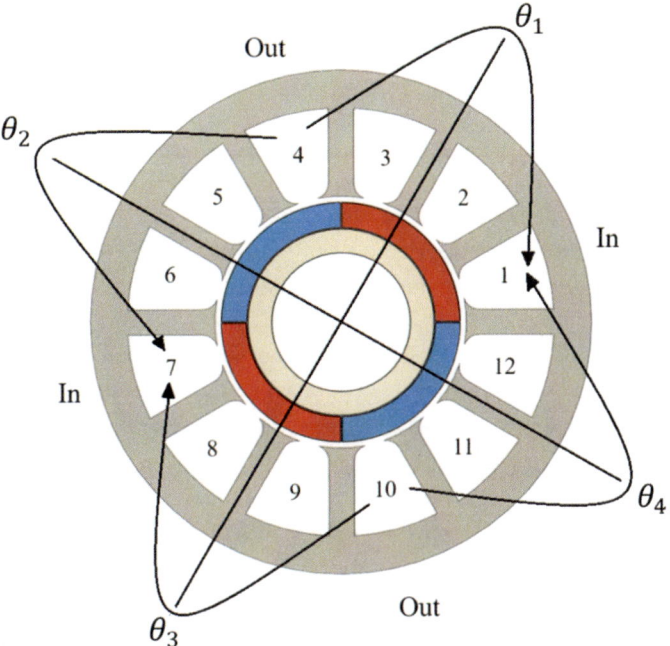

Fig. 3.2 A slot pitch with 4 pole and 12 slot motor

The selection process involves identifying the best possible placements for the coils to ensure that they approximate the ideal angular separations. Once potential coil locations are identified, the appropriate number of coils per phase is then determined, adhering to the initial winding assumptions outlined for the motor design. This step is crucial for ensuring that each phase of the motor is balanced, contributing to optimal motor function and efficiency. By carefully choosing and arranging the coils in this manner, fractional slot motors can achieve performance characteristics that closely mirror those of more ideally configured systems, despite the inherent limitations of their slot and pole configurations.

According to the assumptions laid out for three-phase motor designs, the number of slots (N_s) is always a multiple of three, ensuring that each phase can be uniformly distributed across the motor's stator. This setup is crucial for achieving a balanced electrical configuration and efficient motor operation. In these designs, each coil occupies two slots, but only fills each slot halfway, effectively making it so that each coil occupies a single slot in terms of space. This arrangement allows for a straightforward calculation of the number of coils needed per phase. The calculation for determining the precise number of coils per phase is captured in Eq. 3.7, which divides the total number of slots by three (the number of phases), thereby ensuring that the coils are distributed evenly across each phase. This methodical approach helps maintain the structural and functional integrity of the motor, optimizing both the magnetic interactions within the system and the overall performance of the motor.

$$N_{cph} = \frac{N_s}{N_{ph}} = \frac{N_s}{3} \qquad (3.7)$$

The value N_{cph} represents the number of coil locations that must be found for each phase, and the coil locations for other phases are discovered by applying the phase offset K_0 twice to the coil locations found for phase A.

To understand the process of determining coil locations within a motor, let's examine the example of a four-pole, fifteen-slot motor as illustrated in Fig. 3.3. In this scenario, assume a nominal coil span of three slots. If we have a coil starting in slot 1 and ending in slot 4, this is positioned at 0°E. Moving to the next potential coil location, a coil spanning from slot 2 to slot 5 would be at a relative angle equivalent to one slot pitch. Mathematically, this angle, denoted as θ, is calculated as $\theta = \theta_s = \left(\frac{N_m}{N_s}\right) \cdot 180°$E or 48°E, where N_m is the number of poles and N_s is the number of slots.

Continuing this pattern, a coil spanning from slots 5–8 would theoretically position itself at a relative angle of $\theta = 4\theta_s = 4 \cdot 48°$E or 192°E. However, if this coil is wound in the opposite direction—where the 'In' slot becomes slot 8 and the 'Out' slot becomes slot 5—the relative angle of this coil effectively shifts. This reversal adjusts the relative angle to 192°E−80°E = 12°E. Therefore, any coil with an 'In' slot at kk can have its relative angle determined by Eq. 3.8, which accounts for such directional changes and positional shifts.

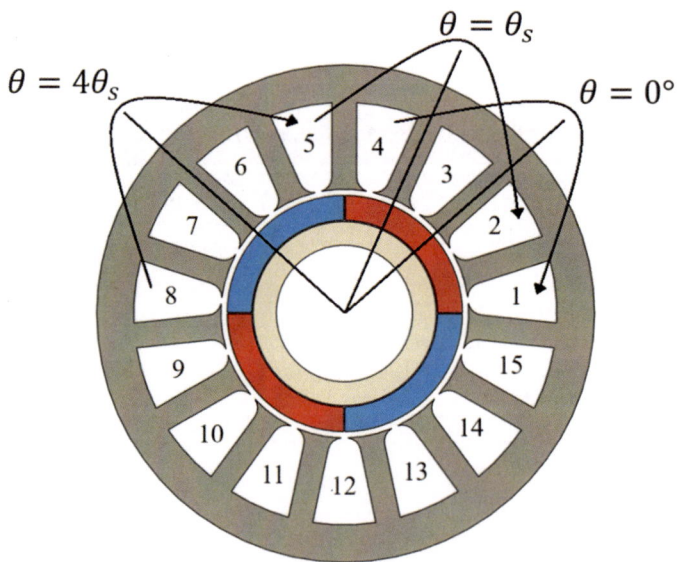

Fig. 3.3 Fifteen slot motor with 4 pole

Table 3.1 In and Out angles of the configuration for the four-pole, fifteen-slot motor in Fig. 3.3

	In/Out						
Coil	1	2	3	4	5	6	7
Angle	0	48	96	144	192	240	288
In	1	2	3	4	5	6	7
Out	4	5	6	7	8	9	10

	In/Out							
Coil	8	9		11	12	13	14	15
Angle	336	384	432	480	528	576	624	672
In	8	9	10	11	12	13	14	15
Out	11	12	13	14	15	1	2	3

$$\theta_c(k) = (k+1)\frac{N_m}{N_s}180°\text{E}. \tag{3.8}$$

This methodical approach to assigning coil positions ensures that the coils are optimally placed to achieve the desired electromagnetic properties and maintain balance within the motor's operation. Each position and direction of winding is calculated to harness the most efficient interaction between the coils and the motor's magnetic field, ensuring effective performance across the motor's operational spectrum.

The In and Out angles and slots remain in the configuration shown in Table 3.1 for the four-pole, fifteen-slot motor in Fig. 3.3.

When calculating coil positions and their relative angles in motor design, it's common to encounter angles that exceed the typical navigational range of $-80°<\theta<180°-80°<\theta<180°$. These angles, while mathematically correct, can be challenging to interpret and apply due to their extension beyond this standard angular range. This complexity often leads to confusion in understanding how coils are physically aligned within the motor structure.

To resolve this issue and simplify the interpretation, it's essential to mathematically adjust these angles so that they fall within the conventional navigational range. This adjustment is done by applying a specific mathematical function, typically a modulo operation or a similar function, as detailed in Eq. 3.8. This function effectively recalibrates any angle exceeding the $180°-180°$ range back into it, ensuring that all calculated angles are expressed within a standard and easily interpretable format. By using this approach, engineers and designers can more straightforwardly determine the precise orientation and placement of coils, facilitating clearer design decisions and more efficient motor configurations. This technique not only aids in the practical layout of motor windings but also helps in enhancing the accuracy of design simulations and the overall assembly process.

$$\theta = \text{rem}(\theta + 180°, 360°) - 180° \tag{3.9}$$

Table 3.2 In and Out angles of the configuration for the four-pole, fifteen-slot motor in Fig. 3.3 considering Eq. 3.8

	In/Out						
Coil	1	2	3	4	5	6	7
Angle	0	48	96	144	−168	−120	−72
In	1	2	3	4	5	6	7
Out	4	5	6	7	8	9	10
	In/Out						
Coil	8	9	10	11	12	13	14	15
Angle	−24	24	72	120	168	−144	−96	−48
In	8	9	10	11	12	13	14	15
Out	11	12	13	14	15	1	2	3

Once determined the value of θ, Table 3.1 can be modified as it is shown in Table 3.2.

In the design of motor windings, when coil angles exceed a magnitude of 90°, a standard practice is to reverse the direction of the coil winding. This reversal effectively shifts the coil angle by 180°, aligning it more suitably within the desired operational range. This technique ensures that the magnetic fields generated by the coils more effectively complement the motor's overall magnetic circuit, enhancing performance and reducing potential interference or inefficiencies that might arise from poorly aligned coils.

Applying this principle to the specific case of a four-pole, fifteen-slot motor, the reversal of coil directions for those coils with angles greater than 90° significantly alters their respective angular placements. This adjustment is a practical step in refining the motor design to ensure optimal function. The results of these adjustments are methodically documented and can be seen in the updated coil data presented in Table 3.3. This table provides a revised and precise overview of the coil configurations, illustrating how such modifications influence the final layout and performance of the motor. By methodically adjusting coil angles and directions, engineers can achieve a more balanced and efficient motor design, tailored to meet specific operational requirements.

When optimizing motor performance, particularly in the selection of coils for phase A, it is beneficial to choose coils that are closest to a 0° angular position while also minimizing the total spread of their angles. This approach aligns the coils more symmetrically around the stator, enhancing the motor's efficiency and balance. In the specific case of a motor with five coils per phase, an analysis of the potential coil positions reveals that coils numbered 1, 5, 8, 9, and 12 are nearest to 0°. The total angular spread among these coils, calculated as the difference between the maximum and minimum angles, is $24° − (−24°) = 48°$.

Selecting these particular coils based on their proximity to 0° and smaller angular spread helps in achieving a more uniform magnetic field distribution, which is crucial for

Table 3.3 In and out angles of the configuration for the four-pole, fifteen-slot motor in Fig. 3.3 considering the correction of 90°

	In/Out						
Coil	1	2	3	4	5	6	7
Angle	0	48	−84	−36	12	60	−72
In	1	2	3	4	5	6	7
Out	4	5	6	7	8	9	10

	In/Out						
Coil	8	9	10	11	12	13	14	15
Angle	−24	24	72	−60	−12	36	84	−48
In	8	9	10	11	12	13	14	15
Out	11	12	13	14	15	1	2	3

reducing torque ripple and improving the motor's smooth operation. The data for these selected coils, once compiled, is sorted by magnitude to further organize and clarify the configuration. This sorted list is then presented in Table 3.4, providing a detailed view of the coil orientations that will be used in the motor. This structured approach to selecting and arranging coils not only optimizes motor performance but also aids in the systematic assembly and documentation of motor design specifications.

To ensure the chosen coil arrangement adheres to all predefined winding assumptions for the motor, a detailed examination is performed, with findings presented in Table 3.5. This table organizes the coils by slot number and applies a coil offset of $K_0 = 10$ slots to align the corresponding coils for phases B and C. The configuration ensures that each slot is fully utilized, containing two coil sides, which affirms the setup as a valid winding according to the motor's design requirements.

In Table 3.5, each row listing two entries confirms that the slotting criteria—each slot hosting two coil sides—are met, establishing a complete and efficient use of space within the motor's stator. This configuration is visually represented in Fig. 3.4, which illustrates the physical placement of phase A coils. Although this particular arrangement lacks angular symmetry, mainly due to the odd number of coils, the directional winding of the coils is

Table 3.4 Detailed view of the coil orientations that will be used in the motor

	In/Out				
Coil	1	5	12	9	8
Angle	0	12	−12	24	−24
In	1	8	15	9	8
Out	4	5	12	12	11

Table 3.5 Coils by slot number considering a coil offset of $K_0 = 10$ for phases B and C

Slot	Phase A	Phase B	Phase C
1	In		Out
2			Out and Out
3		In and In	
4	Out		
5	Out		
6		Out	
7		Out and Out	
8	In and In		
9	In		Out
10		In	Out
11	Out	In	
12	Out and Out		
13			In and In
14		Out	In
15		Out	

logically implemented. Coils 1, 8, and 9 are wound in one direction suitable for one rotor magnet polarity, whereas coils 5 and 12 are wound in the opposite direction to align with the opposing rotor magnet polarity. This strategic placement and winding direction ensure that the electromagnetic forces are optimized across the motor, enhancing performance by maintaining consistent torque and reducing potential magnetic interference.

Consider the example of a ten-pole, twelve-slot motor as depicted in Fig. 3.5, which features a nominal coil span $S^* = 1$ and a phase offset $K_0 = 8$. For this specific motor configuration, the angular slot pitch, θ_s, is calculated using the formula $\theta_s = (N_s/N_m) \cdot 180°\mathrm{E}$, resulting in an angular slot pitch of $150°\mathrm{E}$. This calculation defines the angular separation between consecutive slots based on the motor's structural configuration.

Using this calculated slot pitch, the possible configurations for coils in phase A can be systematically determined. The principal angles for these coils, which are crucial for ensuring optimal electromagnetic interactions within the motor, are meticulously detailed in Table 3.6. This table provides a comprehensive overview of the coils' angular positions relative to a fixed reference point, typically the zero-degree mark. Each coil's position is crucial for aligning magnetic fields properly and achieving the desired mechanical and electrical performance. This arrangement is especially important in motors with a high number of poles, as the interaction between the magnetic fields and the rotor becomes more complex.

Fig. 3.4 Winding layout of phase A for a motor

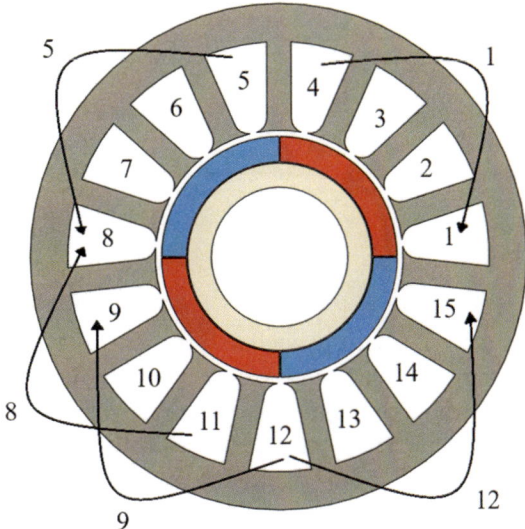

Fig. 3.5. 12 slot motor with 10 pole

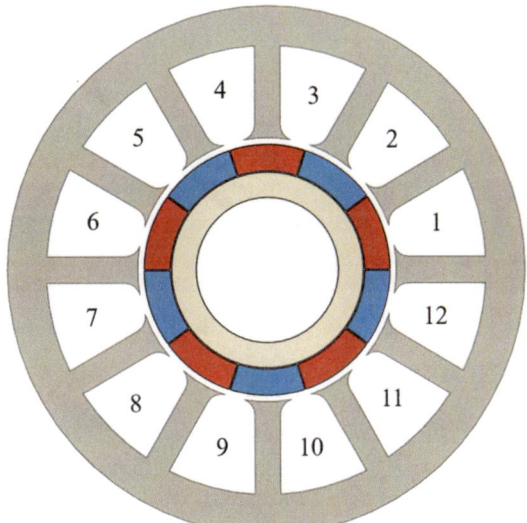

Table 3.6 Overview of the coils' angular positions relative to a fixed reference point

	In/Out											
Coil	1	2	3	4	5	6	7	8	9	10	11	12
Angle	0	150	−60	90	−120	30	−180	−30	120	−90	60	−150
In	1	2	3	4	5	6	7	8	9	10	11	12
Out	2	3	4	5	6	7	8	9	10	11	12	1

Table 3.7 Configuration to maximize efficiency and performance under the given motor specifications

	In/Out											
Coil	1	2	3	4	5	6	7	8	9	10	11	12
Angle	0	−30	−60	90	60	30	0	−30	−60	−90	60	30
In	1	3	3	4	6	6	8	8	10	10	11	1
Out	2	2	4	5	5	7	9	9	9	11	12	12

The set of all possible coils for phase A listed in Table 3.7 reflects the strategic placement necessary to maximize efficiency and performance under the given motor specifications. These configurations consider the unique challenges posed by the ten-pole, twelve-slot design, ensuring that each coil contributes effectively to the motor's overall functionality.

For the ten pole, twelve slot motor being considered, there are four coils per phase. To establish a valid winding configuration, four specific coils need to be selected from the listed options in the table, based on their angular positions to optimize motor performance. Given the critical nature of angular alignment for efficient motor operation, it is evident that coils 1 and 7, which are positioned at 0°, are optimal choices for inclusion due to their direct alignment with the magnetic field's reference direction.

To minimize the angular spread and ensure a balanced distribution of electromagnetic forces within the motor, the next step involves choosing between the two coils positioned at 30° and the two at −30°. Since both of these choices yield a coil arrangement with the same angular spread of 60° (from −30° to 30°), selecting either set would result in comparable motor performance. Opting for the two 0° coils along with the two −30° coils provides a coherent and effective setup, as detailed in Table 3.8. This table confirms that the winding layout for all three phases is valid, with each row containing two entries, thus indicating that each slot is fully utilized with two coil sides.

Figure 3.6 visually represents this winding layout, particularly highlighting the phase A winding. Unlike scenarios with an odd number of coils where angular symmetry might be compromised, this configuration, with an even number of coils, exhibits perfect angular symmetry. The winding is organized into two groups of two coils each, symmetrically positioned on opposite sides of the stator. Within each group, one coil is at 0°, and the other is at −30°, effectively balancing the magnetic interactions across the motor's structure. This symmetrical arrangement not only enhances the aesthetic appeal but also plays a critical role in ensuring the motor operates smoothly with minimal vibration and optimal efficiency.

The winding layout documented in Table 3.8 is strategically designed to minimize the angular spread among the coils that constitute the phase windings, which is typically beneficial for maintaining consistent electromagnetic performance. However, there are several other feasible configurations for winding the ten pole, twelve slot motor, each presenting

Table 3.8 Configuration for a ten pole, twelve slot motor

Slot	Phase A	Phase B	Phase C
1	In		Out
2	Out and Out		
3	In	Out	
4		In and In	
5		Out	In
6			Out and Out
7	Out		In
8	In and In		
9	Out	In	
10		Out and Out	
11		In	Out
12			In and In

Fig. 3.6 Winding layout of phase A for a 12-slot motor

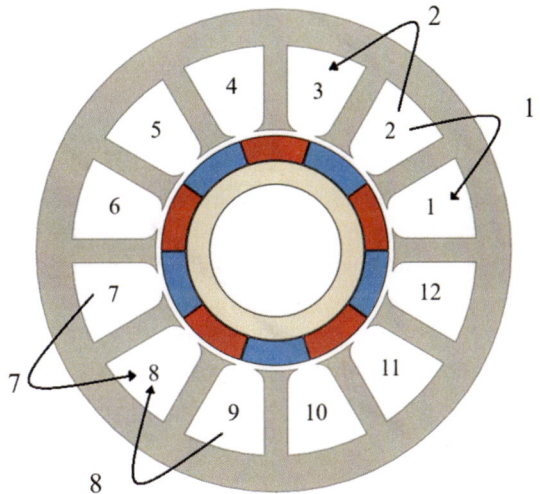

different angular spreads and potential advantages. For instance, another possible config-uration could include selecting the two 0° coils along with one −30° coil and one 30° coil. This arrangement results in a winding with an angular spread of 60°, which, while broader than the minimal spread, might suit specific performance criteria.

Additional alternatives that also result in a 60° angular coil spread include configura-tions such as combining the two 0° coils with the two −60° coils, or the two 0° coils with the two 60° coils. Each of these configurations offers unique characteristics in terms of magnetic field distribution and resultant electromagnetic effects. It is important to note,

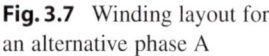

Fig. 3.7 Winding layout for an alternative phase A

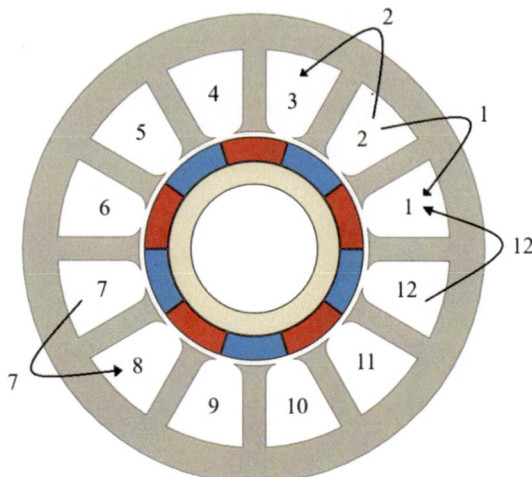

however, that not all potential combinations will result in valid windings. Some configurations may lead to imbalances where some slots might be underfilled while others are overfilled, impacting the motor's efficiency and function.

An example of an alternative valid winding is illustrated in Fig. 3.7. This particular winding includes coils 1, 2, 7, and 12. The choice of these specific coils results in a different angular spread compared to the one shown in Fig. 3.6, which can affect the flux linkage and the back electromotive force (EMF) generated by the motor. Unlike the configuration in Fig. 3.6, these alternative lacks symmetrical placement of coils around the stator, which could influence the motor's vibration patterns and operational smoothness. Understanding these differences is crucial for motor designers who must balance between optimal electromagnetic properties and practical mechanical design requirements to achieve the desired motor performance.

Configuration of the winding procedure

The process for determining valid winding layouts in motor design, as outlined at the start of this chapter and illustrated through subsequent examples, can be systematically approached by following these steps:

1. **Identify the Phase Offset(K_0):** Utilize equations previously defined to find the phase offset K_0. This step is crucial as a valid K_0 is indicative of the possibility to achieve a balanced winding. If no valid K_0 can be identified, it indicates that the motor cannot support a balanced winding with the given pole and slot configuration.

2. **Determine the Nominal Coil Span** (S^*): Use Eq. Eq. (3.2) to calculate the nominal coil span S^*. Adjust this value as necessary to suit specific design requirements or constraints, and denote the final used value as S.

3. **Calculate Coils Per Phase** (N_{cph}): Employ Eq. (3.7) to ascertain the number of coils per phase, ensuring that the total count fits within the motor's structural limits.

4. **Placement of the First Coil in Phase A:** Position the first coil of phase A such that it enters at slot 1 and exits at slot $S + 1$, where S is the designated coil span established in step 2.

5. **Determine Angular Offsets for All Potential Coils:** For each potential coil configuration having a span S, calculate the angular offset relative to the first coil. Express this offset within the principal angle range of $-180°E$ to $180°E$.

6. **Adjust Coil Direction for Excessive Offsets:** For any coil whose angular offset magnitude exceeds $90°E$, reverse the winding directio After reversing, modify the angular offset by $180°E$ to accurately represent the effects of the coil reversal in the winding layout.

7. **Select Coils to Minimize Angular Spread:** From the list of coils and their modified angular offsets, select N_{cph} coils that offer the smallest angular spread among them. It's essential to identify the min.mal spread to enhance the motor's performance and balance. Note that there may be multiple valid and invalid solutions. Carefully choose the coils to ensure a valid and optimally functioning motor.

8. **Determine Winding Layout for Phases B and C:** Once the coils for phase A are selected, use the calculated coil offset K_0 to align the windings for phases B and C. Specifically, shift the phase B winding K_0 slots from the position of the phase A winding. Similarly, adjust the phase C winding another K_0 slots beyond the positioning of phase B. This systematic shifting ensures that each phase is appropriately staggered to maintain the electrical and mechanical balance of the motor.

9. **Validate the Winding Configuration:** After establishing the winding layout for all three phases, verify the validity of the entire winding configuration. A valid winding must fill all slots with exactly two coil sides each—no more, no less. This criterion ensures that the motor's magnetic field is uniformly distributed, promoting efficient operation and minimizing potential mechanical stress or electromagnetic interference. If any slot does not contain exactly two coil sides, or if there are any other discrepcies, the winding configuration is considered invalid, and alternative configurations must be explored. Return to step 7 and reevaluate the selection of coils, possibly choosing different combinations that might yield a valid and more optimal winding layout.

Following these steps ensures a methodical approach to motor winding configuration, maximizing the likelihood of achieving a well-balanced, high-performing motor. Each step builds upon the previous, ensuring that the entire system is harmoniously integrated, from individual coil placement to the holistic interaction of all three phases.

3.5   Connections of the Coils in the Motor

Once the winding layout for a motor has been established, the next critical step involves connecting the individual coils to form the phase windings. The simplest and most common method for achieving this is by connecting all coils in series [5]. This method allows the winding to commence at any slot determined by the winding layout and end at any other slot, provided that all coils are wound in the specified direction and placed in the designated slots.

In addition to a series connection, as explored in this chapter, coils can also be connected in various combinations of series and parallel arrangements to form the phase windings. This flexibility allows for customization based on specific performance requirements or design constraints. However, when connecting coils in parallel, it is crucial to ensure that the back electromotive forces (EMFs) of these coils are identical in amplitude, shape, and phase angle. If these parameters are not perfectly matched, undesirable circulating currents may occur among the coils connected in parallel, which can significantly degrade the motor's performance.

To optimize the motor design, it is necessary to carefully identify combinations of series and parallel connections that avoid issues of circulating currents. This involves meticulous planning and verification of coil characteristics to ensure compatibility before finalizing the connections. By strategically selecting the right combination of series and parallel connections and ensuring coil uniformity, designers can effectively maximize motor efficiency and performance, avoiding potential pitfalls that could lead to suboptimal operation. This methodical approach in connecting the coils is essential for achieving a robust and reliable motor design.

Consider the winding configuration of a four-pole, fifteen-slot motor as shown in Fig. 3.4. In this particular layout, each of the coils possesses a unique angular orientation, meaning no two coils share the same angle. Additionally, given that the number of coils per phase (N_{cph}) is five—a prime number—this constrains the possible configurations for connecting these coils: they must either be connected all in series or all in parallel. Intermediate combinations, involving partial series and partial parallel connections, are not viable due to the prime nature of the coil count.

In this scenario, connecting all coils in parallel would not be advisable. The differing relative angles of the coils would inevitably lead to circulating currents among them, which can adversely affect the motor's efficiency and performance by creating uneven magnetic fields and potential power losses. Therefore, the most feasible and effective method for this specific pole and slot combination is to connect all the coils in series. This approach ensures that the electrical current flows sequentially through each coil, maintaining a consistent torque output and minimizing potential electrical discrepancies. Connecting all coils in series thus emerges as the only reasonable possibility for optimizing the motor's performance given the unique setup of the coils in terms of their angular diversity and the prime number of coils per phase.

In contrast to the previous example, the winding configuration of the ten-pole, twelve-slot motor depicted in Fig. 3.6 presents a different scenario with visible structural groupings. This motor features two distinct coil groups, and notably, the number of coils per phase (N_{cph}) is four, which is not a prime number. This factor allows for more flexibility in how the coils can be connected within the motor.

Each of these coil groups consists of coils positioned at relative angles of $0°$ and $-30°$. An effective strategy here is to connect the coils within each group in series. This means linking one coil at $0°$ directly with another at $-30°$, thus ensuring that the current flows sequentially from one to the next within the same group. Given the coherent angular relationship within each group, connecting the coils in series within the group will align their electromagnetic forces constructively.

Following the series connection within each group, the two groups themselves can then be connected in parallel. This method is advantageous because it equalizes the net back electromotive force (EMF) generated by each group, ensuring that they contribute evenly to the motor's overall operation. Since the back EMFs are aligned and identical in amplitude and phase across the groups, this configuration avoids the creation of circulating currents which could otherwise undermine efficiency and performance.

The adoption of this series within groups and parallel between groups connection scheme effectively forms two parallel paths through the motor's winding. This setup not only optimizes the distribution of electrical current through the motor but also enhances the overall efficiency and stability of the motor's operation by harmonizing the electromagnetic interactions across the coils.

The alternate winding configuration depicted for the ten-pole, twelve-slot motor in Fig. 3.7 illustrates a scenario where no distinct coil groups are formed. This layout significantly restricts the connection options available for the motor's coils. Without groupings that share similar electromagnetic characteristics, the only feasible way to connect the coils is to link them all in series. Attempting to connect these coils in any other configuration, such as in parallel or in mixed groups, would inevitably lead to circulating currents due to the differing back electromotive forces (EMFs) among the coils, which would degrade the motor's efficiency and performance.

This example underscores a broader principle in motor design: the feasibility of connecting all coils in parallel is generally limited to integral slot motors where all coils share the same relative angle, such as shown in the four-pole, twelve-slot motor in Fig. 2.12. Such configurations ensure uniform EMFs across all coils, making parallel connections viable. Additionally, the possibility of arranging coils in combinations of series and parallel connections is influenced not only by the primeness of the number of coils per phase (N_{cph}) but also by the diversity of unique coil offset angles within the winding.

In situations where distinct coil groupings can be clearly identified, as was possible in the initial ten-pole example, some combinations of series and parallel connections become feasible. These arrangements allow for more complex and potentially more efficient wiring configurations, provided they are carefully designed to maintain electromagnetic compatibility among all connected groups. Thus, the structural layout of the coils within the motor significantly impacts the connection strategies that can be effectively implemented.

If the effects of circulating currents are overlooked when connecting coils with different offset angles in parallel, the prime factors of the number of coils per phase (N_{cph}) become crucial in determining feasible combinations for series and parallel connections. For instance, consider the four-pole, fifteen-slot motor configuration, where each phase comprises five coils. The number five has only two prime factors: one and five itself. This indicates that the viable connection configurations include connecting all five coils in series or connecting each coil individually in parallel.

Similarly, in the four-pole, twelve-slot motor, there are four coils per phase. The number four has three prime factors: one, two, and four. This factorization provides additional flexibility in how the coils can be configured: all four coils can be connected in series, each coil can operate independently in parallel (one), or two coils can be connected in series with two such pairs then connected in parallel (two). This flexibility allows for various wiring schemes that can be tailored to specific performance or design requirements.

Understanding the prime factors of N_{cph} is essential for designing efficient and functional motor windings, as it directly influences the electrical characteristics and performance of the motor. Properly aligning the coils based on these factors can help optimize the motor's electromagnetic interactions and overall efficiency, provided that the issues associated with circulating currents are managed or mitigated.

3.6   Determination and Discussion of the Winding Factor

In traditional motor design, the concepts of pitch and distribution factors play crucial roles in understanding and optimizing the back electromotive force (EMF) produced by the motor's coils. The distribution factor accounts for the spatial distribution of coil offset angles across the motor, influencing how the magnetic fields from different coils interact with each other. Conversely, the pitch factor considers the effects of the coil pitch—the distance between the coils—and how this spacing impacts the magnetic flux linkage and the resultant back EMF [6].

When calculating the overall back EMF for a phase winding, both factors are applied to the EMF generated by a coil with a full pitch. By incorporating these factors, designers can derive the net back EMF for the entire phase winding from the EMFs of individual coils. Historically, the calculation of these factors was based on the assumption that the coil back EMF was either sinusoidal or that the effects of harmonics could be disregarded. This assumption was typically valid under conditions where sinusoidal currents were used to drive the motor, simplifying the computation of these factors.

Moreover, in many practical applications, the pitch and distribution factors were often merged into a single metric known as the winding factor. This combined factor simplifies the design and analysis process by providing a consolidated measure that encapsulates both the effects of coil placement and spacing on the motor's performance. By understanding and applying the winding factor, motor designers can effectively predict and enhance the electromagnetic efficiency of the motor, optimizing it for its intended application while maintaining simplicity in the design calculations.

In modern motor design, the traditional manual calculations involving pitch and distribution factors are increasingly obsolete, largely due to advancements in computational tools and software. Motor designers now routinely utilize computer software that can manipulate Fourier series to model and analyze complex waveforms, extending beyond the basic sinusoidal waveforms. This capability allows for the inclusion of an arbitrary number of harmonics in the analysis, providing a more comprehensive understanding of the motor's electromagnetic behaviors.

Consequently, the traditional necessity to derive or apply pitch and distribution factors as separate components in motor design has diminished. Instead, these factors are effectively integrated into more sophisticated computational analyses. However, the principles underlying these factors still significantly influence motor performance. For instance, understanding the spatial arrangement of coils through their offset angles remains crucial. This understanding helps determine what this chapter refers to as the winding factor, which effectively combines the concepts of both pitch and distribution factors where applicable, though explicitly deriving a pitch factor becomes unnecessary.

The use of Fourier series, as outlined in Eq. 3.10, facilitates detailed descriptions of the back EMF generated by each coil within a phase winding. This approach provides a nuanced view of how individual coils contribute to the overall electromagnetic properties of the motor, allowing designers to optimize the motor's performance based on precise mathematical models rather than simplified traditional calculations. This shift not only streamlines the design process but also enhances the accuracy and efficiency of motor designs in contemporary applications.

$$e_k(\theta) = \sum_{n=-\infty}^{\infty} E_n e^{jn(\theta-\theta_k)} \tag{3.10}$$

In the given equation, where θ is expressed in electrical units, the Fourier series coefficients E_nEn, alongside the unit imaginary number j, and the relative angular offset θ_k of each k-th coil, come together to provide a sophisticated mathematical representation of the back electromotive force (EMF) in motor windings. Here, $\theta_1 = 0°$ is established as the baseline or reference angle, ensuring consistency in how angular offsets are calculated relative to this initial position.

This mathematical expression delineates that while each coil's back EMF may differ in terms of phase due to the angular offset θ_k, their amplitudes and waveform shapes remain consistent across the board. Such uniformity in amplitude and shape, despite differences in phase offset, aligns perfectly with the winding assumptions laid out earlier. These assumptions presume that aside from their spatial positioning (reflected in phase differences), the coils' electromagnetic properties are otherwise equivalent—this equivalence is crucial for achieving a balanced and efficient motor operation.

The implication of this approach is significant for motor design, as it simplifies the analysis and synthesis the motor's electromagnetic characteristics. By acknowledging that all variations in coil back EMFs are purely due to differences in angular positioning, designers can focus on optimizing phase alignment and distribution without concern for inconsistencies in other electromagnetic properties of the coils. This leads to more streamlined and effective motor designs, where the primary variable manipulated for performance enhancement is the angular offset rather than altering the intrinsic properties of each coil.

If N_{cph} coils are connected in series, the back EMFs of the individual coils combine to form the phase back EMF. This cumulative effect is mathematically modeled as described in Eq. 3.11.

$$e_{ph}(\theta) = \sum_{k=1}^{N_{cph}} e_k(\theta) \tag{3.11}$$

Combining Eqs. 3.10 and 3.11, the final model can be modeled as follows:

$$e_{ph}(\theta) = \sum_{k=1}^{N_{cph}} \sum_{n=-\infty}^{\infty} E_n e^{jn(\theta-\theta_k)} = N_{cph} \sum_{n=-\infty}^{\infty} K_{wn} e^{jn(\theta-\theta_k)}$$

$$K_{wn} = \frac{1}{N_{cph}} \sum_{k=1}^{N_{cph}} e^{-jn\theta_k} \tag{3.12}$$

In the specific scenario where all coil offset angles, θ_k, are zero, as typically observed in integral slot motors, a perfect alignment is achieved. An example of this is the four-pole, twelve-slot motor configuration. Under these conditions, the winding factor, K_{wn}, is equal to one for all harmonic indices nn. This uniformity indicates that the phase back EMF is not influenced by angular discrepancies among the coils because there are none. Consequently, the phase back EMF, as described by Eq. 3.12, becomes a direct amplitude-scaled replica of the individual coil back EMFs. This means that the collective output of the phase back EMF mirrors the individual coil EMFs in both shape and waveform, merely adjusted in amplitude. This alignment maximizes the efficiency of electromagnetic induction within the motor, ensuring that the power output is optimal without any phase shift distortions or reductions in the EMF efficiency due to angular misalignment of the coils.

When the coil offset angles θ_k vary, meaning they are not uniformly zero, the winding factor as outlined in Eq. (3.12) plays a crucial role in determining how the harmonics of the individual coil back EMFs influence the overall harmonics of the phase back EMF. This winding factor quantifies the effect of the angular distribution of coils on the harmonic content of the motor's electromagnetic output. Depending on the specific winding layout, certain harmonics present in the individual coil back EMFs may be significantly reduced or entirely absent in the phase back EMF.

The impact of varying winding layouts on the harmonics is illustrated through Figs. 3.8, 3.9 and 3.10, which depict the amplitude of winding factors across different harmonic indices nn for the three winding layouts examined earlier in this chapter. Typically, ideal back EMFs do not contain even harmonics due to their inherent half-wave symmetry, rendering winding factors for even harmonic indices irrelevant. Comparing Figs. 3.9 and 3.10, the alternative winding layout shown in Fig. 3.7 does not significan.ly alter the fundamental winding factor ($n=1$) compared to the original layout in Fig. 3.6. Both configurations maintain a similar level at this fundamental harmonic.

However, a more detailed analysis of other harmonics reveals that the winding factors in the original layout (Fig. 3.6) generally exhibit greater amplitudes. This suggests that the original winding might produce a back EMF with more pronounced harmonic content. Conversely, the alternative layout in Fig. 3.7, with its comparably lower harmonic amplitudes, might result in a more sinusoidal back EMF. This reduction in harmonic content can be advantageous, as a more sinusoidal back EMF typically contributes to smoother motor operation and reduced electromagnetic interference. Thus, depending on the application requirements, selecting a winding layout that optimizes the sinusoidal of the back EMF could be beneficial for enhancing motor performance and efficiency.

Fig. 3.8 Winding factor for a slot motor of Fig. 3.4

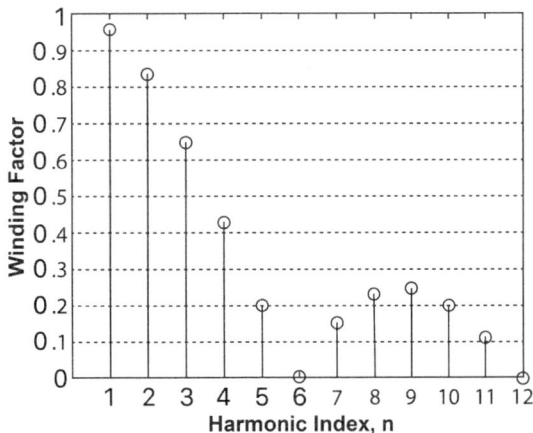

Fig. 3.9 Winding factor for a slot motor of Fig. 3.6

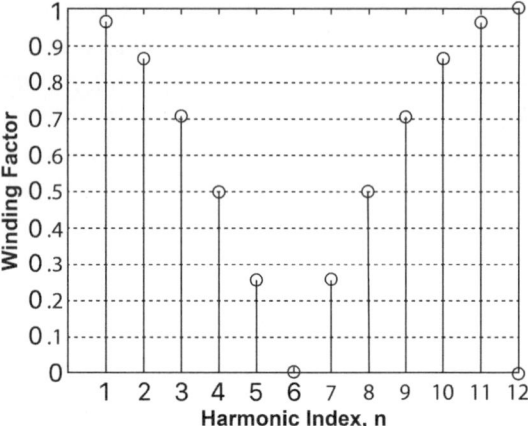

Fig. 3.10 Winding factor for a slot motor of Fig. 3.7

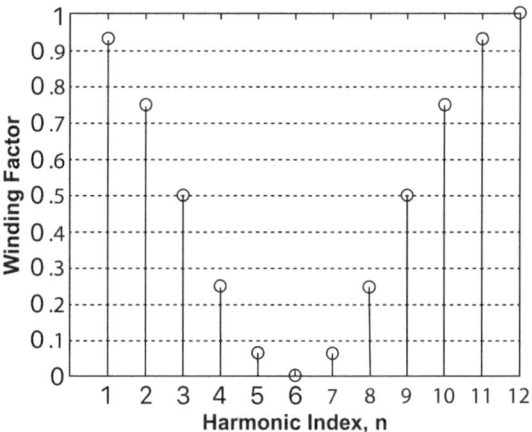

3.7   Inductance

As we explore the concepts of winding configurations and their impact on motor performance throughout this chapter, it becomes evident that traditional methods for computing air gap inductance and slot leakage inductance, as previously discussed, do not directly apply to the scenarios presented here. The focus so far has been primarily on full pitch windings within integral slot motors. However, the broader and more complex configurations explored in this chapter necessitate a reevaluation and further investigation into how inductance is computed in these varied setups [7] (Fig. 3.11).

To address the complexities of calculating inductance, flux linkage, and back EMF for a general winding layout, one effective approach is to decompose the winding into

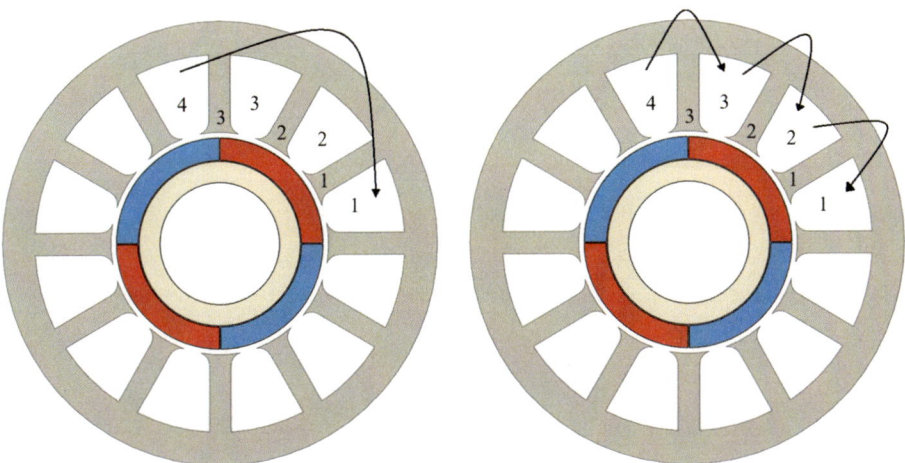

Fig. 3.11 Coil equivalence for a single tooth

a series of single tooth coils. This method is exemplified in Fig. 6.11, where a comparative analysis between different motor cross sections is provided. In this illustration, the winding configuration in the left motor cross section effectively mirrors the electromagnetic properties—air gap inductance, flux linkage, and back EMF—of the three coils series-connected as shown in the right motor cross section.

This equivalence is maintained through the principle of net ampere-turns (*NI*) cancellation in intermediate slots, such as slots 2 and 3 in the figure. In these slots, one coil side introduces current into the cross section while another extracts it, effectively neutralizing the magnetic contributions of these slots. Consequently, the only significant contributions come from the current entering at slot 1 and exiting at slot 4, mirroring the single tooth coil configuration. This method not only simplifies the analysis but also ensures a more precise calculation of inductance in motor designs that deviate from the standard full pitch, integral slot configurations. This approach facilitates a deeper understanding of the magnetic interactions within more complex motor structures and contributes to more accurate and efficient motor design practices.

To effectively compute air gap inductance, let's examine a practical example using the four-pole, twelve-slot motor. Figure 6.12 illustrates this scenario, where the left side of the figure displays the actual phase A winding configuration, and the right side shows its equivalent in terms of a single tooth coil. This comparative layout helps clarify the transformation from a more complex arrangement to a simplified model that accurately reflects the magnetic interactions occurring in the motor.

The calculation of air gap inductance in this context involves applying the principles outlined in Eq. 3.13. This equation integrates the specific geometrical and magnetic properties of the motor's design to derive the inductance values. By focusing on the single

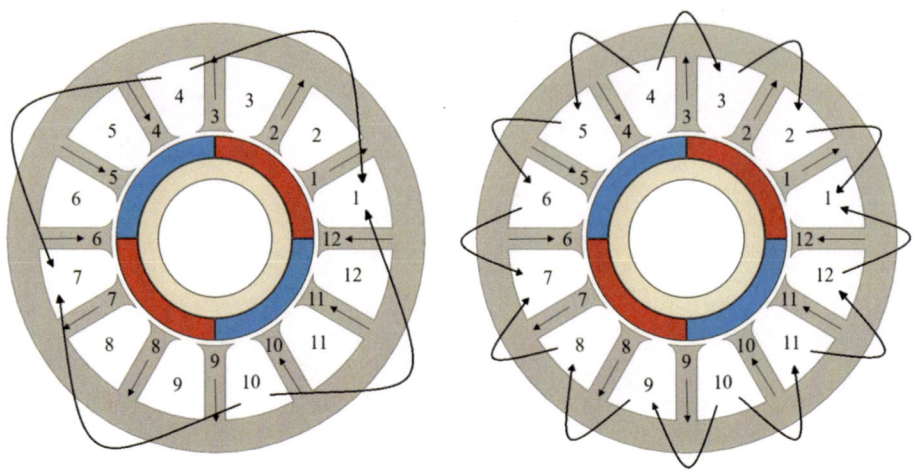

Fig. 3.12 Winding for the phase A for a 12 slot motor

tooth coil equivalent, as shown on the right side of Fig. 6.12, the equation simplifies the process of understanding how magnetic fields interact across the air gap of the motor. This approach provides a clear, mathematical pathway to measure the inductance, taking into account the direct effects of the magnetic fields generated by the current flowing through the actual windings.

$$
L_g = \frac{2\pi \mu_0 L_{st} R_{r0}}{1 + \frac{l_m}{\mu_R C_\phi}} \cdot N^2
$$
(3.13)

Using this method not only streamlines the inductance calculation but also enhances the accuracy of the design and analysis of the motor by isolating the key factors that influence electromagnetic behavior in the air gap. This targeted approach allows for more precise engineering adjustments and optimizations in motor design, leading to improved performance and efficiency (Fig. 3.13).

The concept of the magnetic circuit for a single tooth coil equivalent is depicted in Fig. 6.13. This figure illustrates how the magnetic properties of the motor are simplified into a model that clearly represents the key elements influencing magnetic behavior within the motor's structure. In this model, R_{gm} represents the combined reluctances of the air gap and the magnet that each tooth of the rotor experiences. Additionally, S_k for $k = 1, 2, ..., N_s$ is (where N_s is the number of slots) acts as both a sign and scale factor. This factor crucially carries the sign of the magnetomotive force (MMF) source for each tooth, as well as indicating the relative number of turns associated with each coil.

For instance, as shown in Fig. 6.12, S_1 is set to $+1$, suggesting a positive direction of the MMF for the first tooth, while S_{12} is -1, indicating an opposite MMF direction for the twelfth tooth. This sign distinction is essential in determining the overall direction

Fig. 3.13 Description of the magnetic circuit of Fig. 3.12

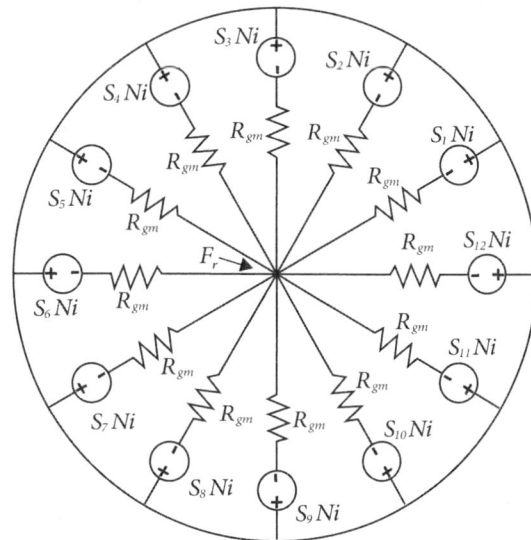

and magnitude of the magnetic flux through each part of the motor. In scenarios where multiple single tooth coils encircle a single tooth, the sign and scale factor for that tooth would be expressed as $S = \pm m$, where mm is the number of coils around the tooth, and the sign (+ or −) is dictated by the direction of the flux through that particular tooth.

This simplified magnetic circuit model allows for a more straightforward calculation and understanding of the overall magnetic behavior of the motor, especially in how the magnetic flux is distributed and interacts within the air gap and across the motor's magnetic elements. It provides a clear framework for analyzing and optimizing motor design from a magnetic perspective, ensuring that all components are appropriately aligned and contributing efficiently to the motor's operation.

To determine the magnetomotive force (MMF) at the center node relative to the stator yoke in an electric motor, one must consider the total flux that exits from the center node. The principle behind this calculation is based on Kirchhoff's law for magnetic circuits, which posits that the algebraic sum of magnetic fluxes around a closed loop must equal zero. In this context, it involves summing the flux leaving the center node and equating this sum to zero to ensure magnetic balance.

This calculation can be effectively modeled using Eq. 3.14. This equation encapsulates the dynamics of the magnetic interactions within the motor by considering contributions from various components of the motor's magnetic circuit. Specifically, it factors in the magnetic fluxes contributed by each tooth or coil that connects to the center node, adjusted for their respective magnetomotive forces and reluctances.

$$\sum_{k=1}^{N_s} \phi_k = \sum_{k=1}^{N_s} \frac{F_r + S_k Ni}{R_{gm}}. \tag{3.14}$$

where N_s is the number of slots, which is equal to the number of teeth. By setting up the equation to sum these flux contributions and requiring that they equal zero, the MMF at the center node can be isolated and calculated. This approach provides a comprehensive view of the magnetic conditions at a critical point in the motor, facilitating a deeper understanding of how magnetic fields distribute across the motor structure. Accurately modeling this interaction is crucial for optimizing motor design to improve efficiency, reduce losses, and ensure stable operation. Solving Eq. 3.14 for the rotor MMF the value of F_r. can be computed as follows:

$$F_r = -\frac{NI}{N_s} \sum_{k=1}^{N_s} S_k \tag{3.15}$$

The net flux linkage (λ) of the winding in an electric motor is fundamentally defined by the relationship $\lambda = N\phi$, where N represents the number of turns in the winding, and ϕ is the summation of the fluxes associated with each tooth, as detailed in Eq. 3.15. This equation captures the magnetic contributions from individual tooth fluxes, considering the directionality or polarity of each contribution as specified by the sign factors S_k. These factors S_k indicate the direction (positive or negative) of the magnetomotive force (MMF) for each tooth, ensuring that the flux contributions are correctly aligned according to their actual influence in the motor's magnetic circuit.

The calculation of λ thus incorporates these fluxes, properly adjusted by their respective signs, to accurately reflect the total magnetic linkage through the winding. The specific computation of λ., taking into account all these elements, is formulated in Eq. 3.15. This equation synthesizes the contributions from each component flux into a cohesive measure of total flux linkage. By applying Eq. 3.16, the value of λ can be precisely calculated, providing critical insights into the efficiency and effectiveness of the magnetic field interactions within the motor. This understanding is vital for optimizing motor design to enhance performance and reduce energy losses in practical applications.

$$\lambda = N \sum_{k=1}^{N_s} \text{sign}(S_k)\phi_k = N \sum_{k=1}^{N_s} \text{sign}(S_k)\left(\frac{F_r + S_k Ni}{R_{gm}}\right). \tag{3.16}$$

where the function sign($\cdot$) determines the sign of its input. Employing the relationship $L = \lambda/i$, and incorporating Eq. 3.16 into Eq. 3.15, the air gap inductance for coils connected in series is calculated using Eq. 3.17.

$$L_g = \frac{\lambda}{i} = \frac{N^2}{R_{gm}} \sum_{k=1}^{N_s} \text{sign}(S_k)\left(S_k - \frac{1}{N_s} \sum_{m=1}^{N_s} S_m\right) \tag{3.17}$$

Assuming the expressions for the air gap and magnet reluctances to express them on a per tooth basis produces the expression defined in Eq. 3.18.

Table 3.9 Scale factors S_k values

Tooth	1	2	3	4	5	6	7	8	9	10	11	12
S_k	1	1	1	−1	−1	−1	1	1	1	−1	−1	−1

$$R_{gm} = R_g + R_m = \frac{g}{\mu_0 A_g} + \frac{l_m}{\mu_R \mu_0 A_m} = \frac{g + \frac{l_m}{\mu_R C_\phi}}{\mu_0 A_g} \tag{3.18}$$

where the flux concentration factor $C_\phi = A_m/A_g$ is applied to streamline the formula. By incorporating the air gap cross-sectional area $A_g = L_{st}\theta_s R_{r0}$, with $\theta_s = 2\pi/N_s$, into Eq. 3.18 and integrating this result into Eq. 3.17, we derive the formula for air gap inductance as specified in Eq. 3.19.

$$L_g = N^2 \frac{2\pi \mu_0 L_{st} R_{r0}}{g + \frac{l_m}{\mu_R C_\phi}} \left(\frac{1}{N_s} \sum_{k=1}^{N_s} \text{sign}(S_k) \cdot \left(S_k - \frac{1}{N_s} \sum_{m=1}^{N_s} S_m \right) \right) \tag{3.19}$$

For the four-pole, twelve-slot motor depicted in Fig. 6.12 and analyzed through the magnetic circuit in Fig. 6.13, the scale factors S_k are specified as the values listed in Table 3.9.

If the value of Eq. 3.19 is used, the gap inductance can be determined by the expression given in Eq. 3.20 (Fig. 3.14).

$$L_g = N^2 \frac{2\pi \mu_0 L_{st} R_{r0}}{g + \frac{l_m}{\mu_R C_\phi}} \tag{3.20}$$

To demonstrate a scenario where multiple coils are positioned around certain teeth, let's examine the four-pole, fifteen-slot motor illustrated in Fig. 6.14. In this particular motor design, the layout involves several coils surrounding individual teeth, highlighting a more complex winding pattern. The scale factors S_k, which are crucial for quantifying the magnetomotive force (MMF) contributed by each coil relative to its position and number of turns, are meticulously outlined for this motor cross-section. These scale factors are defined and listed as shown in Table 3.10. This table provides a detailed breakdown of the scale factors, ensuring that each coil's contribution to the overall magnetic field of the motor is accurately represented and calculated, crucial for analyzing and optimizing the motor's electromagnetic performance.

Utilizing the information provided in Table 3.10, the air gap inductance for the four-pole, fifteen-slot motor, with all coils connected in series, is precisely calculated. The specified data helps in accurately determining the contribution of each coil to the motor's overall magnetic circuit, which is essential for this computation. The air gap inductance is a critical parameter for assessing the efficiency and performance of the motor, particularly in terms of its electromagnetic interactions. The formula for this calculation is comprehensively detailed in Eq. 3.21, which integrates the scale factors and other relevant

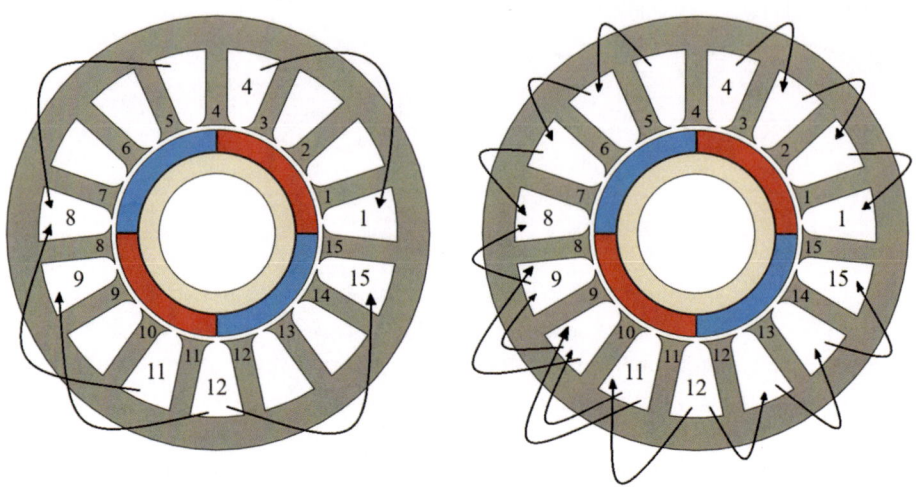

Fig. 3.14 Winding of phase A for a 15-slot motor

Table 3.10. Scale factors S_k values for more than one coil

Tooth	1	2	3	4	5	6	7
S_k	1	1	1	0	-1	-1	-1
Tooth	8	9	10	11	12	13	14	15
S_k	1	2	2	1	-1	-1	-1	0

variables from the table. This equation ensures that the inductance is quantified based on the current configuration of the motor, providing a solid foundation for further analysis and optimization of the motors design.

$$L_g = N^2 \frac{2\pi \mu_0 L_{st} R_{r0}}{g + \frac{l_m}{\mu_R C_\phi}} \cdot \frac{14.8}{15} \tag{3.21}$$

In conclusion, the calculation of air gap inductance as delineated by Eq. 3.19 hinges critically on the determination of tooth scale factors. These factors are integral to understanding the contribution of each tooth to the motor's overall magnetic circuit. Identifying these scale factors becomes straightforward once a valid winding layout has been established. The relative simplicity of Eq. 3.19 is largely attributed to the concept of single tooth coil equivalence. This model simplifies the otherwise complex reality of interacting magnetic fields within the motor by equating the effects of a series of coils to that of a single coil per tooth. Without this conceptual simplification, the process of computing air gap inductance would involve much more intricate calculations, dealing with the cumulative effects of multiple, interacting coils, which would significantly complicate both the analysis and the design process of electric motors.

3.8   Leakage Inductance of Slots

Integral slot motors featuring full pitch windings, exemplified by the four-pole, twelve-slot motor illustrated in Figs. 6.2 and 6.12, consistently have two coil sides per slot fully occupied by a phase winding. In such configurations, the calculation of slot leakage inductance is typically addressed per slot. This setup ensures that the electromagnetic interactions are contained within each slot, simplifying the inductance calculations [8].

Conversely, fractional slot motors like the four-pole, fifteen-slot motor displayed in Figs. 6.4 and 6.14 present a different scenario where not every slot contains windings from just one phase. In cases where slots house coil sides from two different phases, there exists what is termed mutual slot leakage inductance. This additional complexity means that the straightforward principles used for calculating self-inductance in integral slot motors do not directly apply. The presence of coil sides from multiple phases in the same slot introduces inter-phase electromagnetic interactions, complicating both the measurement and the management of leakage inductance within the motor (Fig. 3.15).

If the mutual inductance component is disregarded, the slot leakage inductance associated with a single coil side that occupies half of a slot can vary significantly depending on its placement within the slot. Specifically, coil sides located at the bottom of the slot typically exhibit higher slot leakage inductance compared to those positioned at the top. This variation is primarily due to differences in the magnetic path lengths and the proximity to the stator yoke. Coil sides at the bottom of the slot are generally farther from

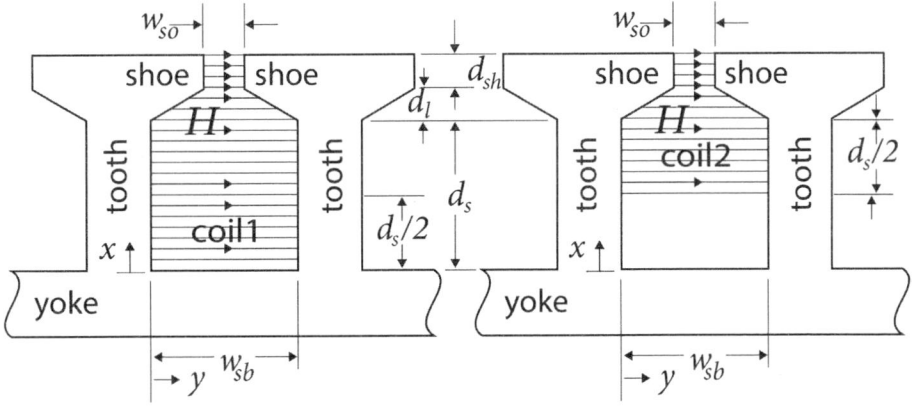

Fig. 3.15 Possibilities for a coil side placement

the slot opening and closer to the yoke, which can increase the magnetic reluctance and, consequently, the leakage inductance. This positioning effect needs to be carefully considered in motor design, as it impacts the overall efficiency and performance of the motor by affecting how effectively the magnetic fields are contained and utilized within the machine.

To better understand the impact of coil side placement on slot leakage inductance, let's examine two different placements as depicted in Fig. 6.15. Specifically, focus on the placement of the coil side at the bottom of the slot, as shown on the left side of the figure. This position significantly affects the magnetic field behavior and the resulting slot leakage inductance. Using the analytical technique described in Chapter 4, we can explore this scenario in depth.

In this bottom-slot placement, the magnetic field intensity, which crosses the slot from one tooth to another, reaches its maximum value not at the slot opening ($x = d_s$), but halfway down the slot ($x = d_s/2$). From this midpoint ($x = d_s/2$) to the bottom of the slot ($x = d_s$), the field intensity remains constant. This constant value, denoted as $F = Ni/w_{sb}$, where Ni represents the product of the number of turns and the current, and w_{sb} is the width of the slot bottom, indicates that the magnetic flux linkage is uniform in this region. Therefore, the inductance contributed by the region from $x = d_s/2$ to $x = d_s$ an be directly calculated using the formula N^2P, where P represents the permeance of that particular slot segment.

.When these concepts are applied, and it is recognized that the number of turns is N (instead of $2N$ which might be used in other calculations involving full slots), the total slot leakage inductance per slot for coils positioned at the slot bottom can be defined by Eq. 3.22. This equation will take into account these specific dynamics and provide a precise calculation of the slot leakage inductance for this configuration, offering essential insights for optimizing motor design based on the physical placement of coil sides within slots.

$$L_{s1} = N^2 \left(\frac{\mu_0 d_s L_{st}}{6 w_{sb}} + \frac{\mu_0 d_s L_{st}}{62} + \frac{\mu_0 d_t L_{st}}{(w_{so} + w_{sb})/2} + \frac{\mu_0 d_{sh} L_{st}}{w_{so}} \right) \qquad (3.22)$$

The first term in the parentheses refers to the coil area inductance, the second term represents the inductance of the area above the coil, and the third and fourth terms denote the taper ratio and slot opening components, as outlined in this chapter.

When a coil side is positioned at the top of the slot, as depicted on the right in Fig. 6.15, the dynamics of the magnetic field interaction within the slot change significantly. In this setup, there is no magnetic field generated by the coil crossing the slot area below the winding. This absence of magnetic field in the lower part of the slot is attributed to the high permeability of the teeth and the stator yoke, which effectively channel all the magnetic flux around the bottom of the slot. The magnetic fields preferentially follow the path of least resistance, offered by the metallic components, rather than crossing their gap in the lower part of the slot where the magnetic resistance is higher.

In this scenario, the equivalent second term in Eq. 3.22, which typically accounts for the inductance of the area above the coil, becomes irrelevant because there is no coil-induced magnetic field in the space below the coil to consider. Conversely, the first term of Eq. 3.22, which concerns the coil area inductance, now specifically applies to the area at the slot top where the coil is located.

Consequently, the slot leakage inductance for a coil side situated at the top of the slot is computed using Eq. 3.23. This equation focuses solely on the contributions from the top part of the slot, reflecting the altered magnetic field distribution and the corresponding inductive effects. This specialized calculation provides a more accurate determination of the slot leakage inductance in scenarios where the coil is not centrally positioned within the slot.

$$L_{s2} = N^2 \left(\frac{\mu_0 d_s L_{st}}{6 w_{sb}} + \frac{\mu_0 d_t L_{st}}{(w_{so} + w_{sb})/2} + \frac{\mu_0 d_{sh} L_{st}}{w_{so}} \right). \tag{3.23}$$

A comparative analysis of Eqs. 3.22 and 3.23 reveals that the slot leakage inductance for coil sides located at the slot bottom is higher than that for coil sides situated at the slot top. This discrepancy arises because the bottom coil sides are more deeply embedded within the slot, surrounded by more magnetic material which enhances their inductive properties. In contrast, coil sides at the slot top are closer to the slot opening where the magnetic field can more easily dissipate, resulting in lower inductance.

Given this variance in inductance based on coil placement, it is advantageous for motor design to balance the number of coil sides in both the slot bottom and slot top across all phase windings. This ensures that each phase has a similar inductance, promoting uniform performance and efficiency. However, if phases are wound sequentially—one after the other—the first phase typically ends up with more coil sides in the slot bottom, while the last phase wound tends to have more at the slot top. This winding sequence can lead to slight discrepancies in inductance between the phases.

Despite these differences, in most cases, the impact on the overall motor performance is minimal. The total phase inductance, which includes contributions from both the air gap and end turn inductances, generally offsets the relatively minor differences between the inductances calculated by Eqs. 3.22 and 3.23. Thus, while the inductive characteristics at different slot positions do affect each phase, the overall effect on the motor's operation is typically small, allowing for effective performance even with this variation in coil placement.

References

1. J. Cros, P. Viarouge, Synthesis of high performance PM motors with concentrated windings. IEEE Trans. Energy Convers. **17**(2), 248–253 (2002)
2. G. Huth, Permanent-magnet-excited AC servo motors in tooth-coil technology. IEEE Trans. Energy Convers. **20**(2), 300–307 (2005)

3. E. Carraro, N. Bianchi, S. Zhang, M. Koch, Design and performance comparison of fractional slot concentrated winding spoke type synchronous motors with different slot-pole combinations. IEEE Trans. Ind. Appl. **54**(3), 2276–2284 (2018)
4. Y. Du, F. Xiao, W. Hua, X. Zhu, M. Cheng, L. Quan, K.T. Chau, Comparison of flux-switching PM motors with different winding configurations using magnetic gearing principle. IEEE Trans. Magn. **52**(5), 1–8 (2015)
5. D.G. Dorrell, A.C. Smith, Calculation of UMP in induction motors with series or parallel winding connections. IEEE Trans. Energy Convers. **9**(2), 304–310 (1994)
6. A. Mohammadpour, A. Gandhi, L. Parsa, Winding factor calculation for analysis of back EMF waveform in air-core permanent magnet linear synchronous motors. IET Electr. Power Appl. **6**(5), 253–259 (2012)
7. T.J.E. Miller, M.I. McGilp, D.A. Staton, J.J. Bremner, Calculation of inductance in permanent-magnet DC motors. IEE Proc. Electr. Power Appl. **146**(2), 129–137 (1999)
8. P. Ponomarev, P. Lindh, J. Pyrhönen, Effect of slot-and-pole combination on the leakage inductance and the performance of tooth-coil permanent-magnet synchronous machines. IEEE Trans. Industr. Electron. **60**(10), 4310–4317 (2012)

2D Sketching in CAD Software for 3D Printing

4

4.1   Background

2D sketching is a fundamental step in obtaining successful 3D models that can reliably reflect the characteristics of an object in a digital representation [1]. Such a 2D sketching would allow the designer to progressively add the distinctive characteristics of the 3D model to a base model. Additionally, this progressive process will ease the visualization of problems during the manufacturing of the object, helping to improve the design in the early steps of the project.

Additionally, as many Computer Assisted Desing softwares (CAD softwares) include modules to simulate the physical properties of the 3D object being modeled [2], it is possible to evaluate the resistance of the model, its mass distribution, as well as the strengths of the piece under load. All of this allows for the improvement of the design at an early stage and, therefore, expedites its final production.

Nevertheless, the use of 3D models is not free of inconveniences. One of them is their capacity to design impossible structures. Given its flexibility, creating a piece that can be impossible to manufacture through traditional techniques is possible. For Example, the CAD software has no "common sense" to detect the problems implied in generating an infinitesimally thin piece [3] or designing a structure with holes in places where it is impossible to introduce a milling tool [4]. In Fig. 4.1, a piece with a hole in its central column is presented to improve the visualization of such complex manufacturing structures. Such a structure is impossible to manufacture with traditional subtractive techniques, requiring a special or additive process like 3D printing. Such a characteristic is due to the lack of space to introduce a milling tool, which restricts the process implemented.

Considering the above, it is convenient to design objects that can be easily fabricated with traditional operations that can be made with the available tools. Thus, a design that considers such restrictions tends to reduce the manufacturing process and the time and

© The Author(s), under exclusive license to Springer Nature Switzerland AG 2025
E. Cuevas et al., *DC Motors*, Synthesis Lectures on Engineering, Science, and
Technology, https://doi.org/10.1007/978-3-031-64354-5_4

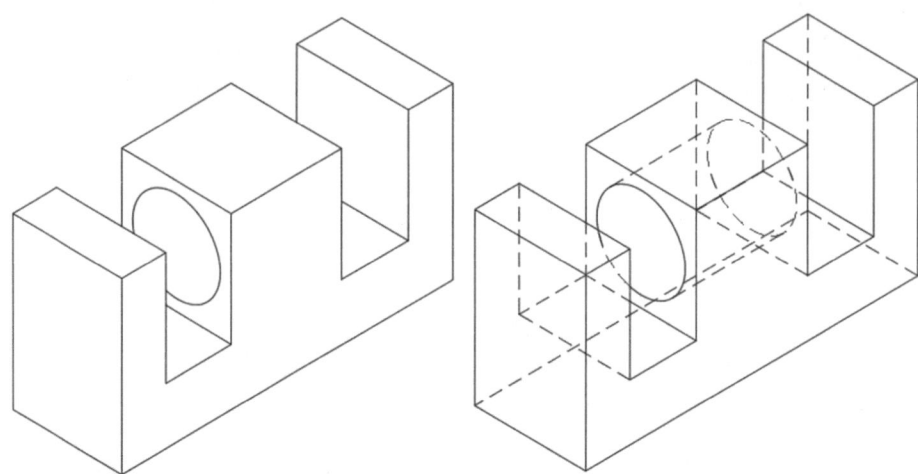

Fig. 4.1 3D model of a complex object challenging to manufacture with traditional subtractive techniques

cost associated with the production of the object, which are desirable characteristics of a design.

Thus, as a response to construction restrictions, most software oriented to CAD design tends to have operations that emulate some traditional tools. This allows for an easy emulation of the manufacturing process and early detection of difficulties in constructing the designs.

Even though there are several CAD softwares options, most of them tend to share the same set of operations. Thus, in the following sections and chapters, this operation will be described, taking as reference the set of instructions of Fusion 360 (Autodesk software). This selection was made considering the ease of use, low cost, and versatility of their tools. However, a crosscheck was made to ensure that the descriptions can be used to understand the design process for other software, such as SolidWorks or other traditional CAD programs.

Finally, as this chapter has the objective of introducing the design oriented to 3D printing, in some operations, some tips & tricks will be added to facilitate the process of 3D printing a model or to avoid some common problems derived from the use of 3D printers.

4.2   2D Sketch Design

A fundamental step of 3D printing consists of the capacity to portray the complexities of a 3D body on a set of 2D sketches, over which an operation will be made to add volume and transform them into a 3D characteristic.

To generate such sketches, it is only required to select the "DESIGN" environment, click on "SOLID," and choose the option "Create Sketch" (Fig. 4.2), after which the software will ask for the plane over which it is desired to work. As a convention in CAD software, it is commonly used a cartesian space with three perpendicular axes, where the "floor" is defined with the axes X (red) and Y (green), and the height is defined with the Z axis (blue), as shown in Fig. 4.3.

Once the plane to work is selected, the software will offer a new set of tools over de SKETCH menu and several restrictions to define the model to be designed. For Example, Fig. 4.4 captures the contextual band available while editing a 2D sketch.

As can be observed, the contextual strip for the design of 2D sketches has several sections, of which three are of main interest. The first, the "CREATE" section, allows the

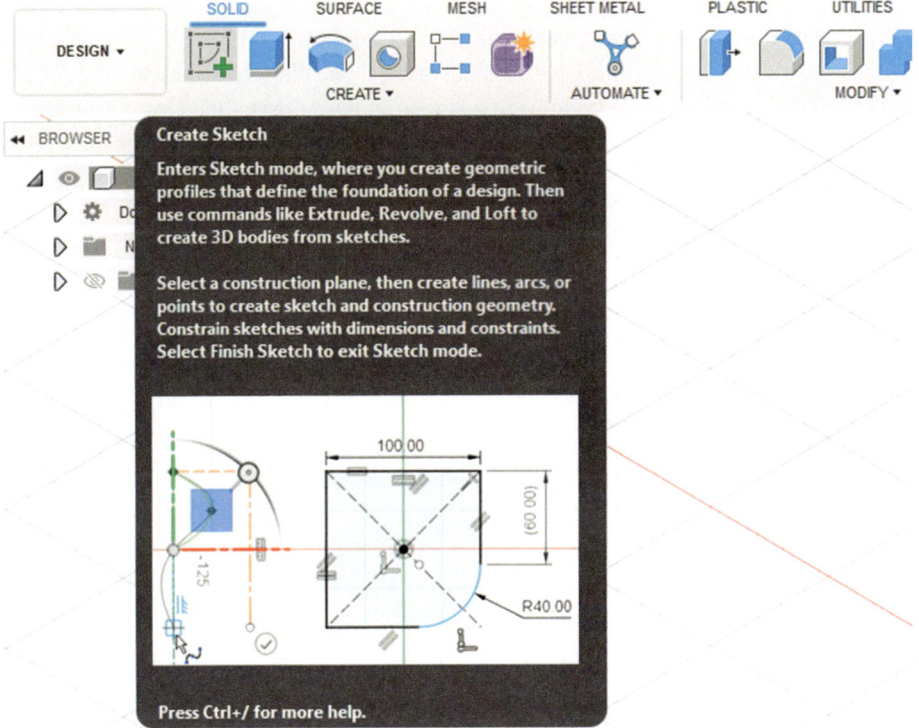

Fig. 4.2 Button "Create Sketch"

Fig. 4.3 Traditional cartesian
system

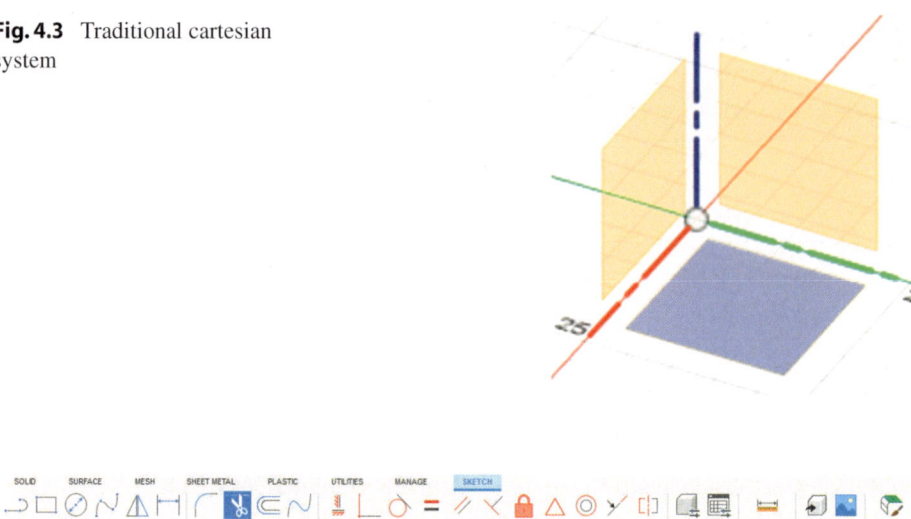

Fig. 4.4 Contextual band with the options to add, modify, and constrain different elements of the
2D sketch

addition of new elements to the created sketch; the second, the "MODIFY" section, offers
a set of tools to modify the available elements of the sketch; while the last section, the
"CONSTRAIN" section, provides tools to add restrictions between the elements drawn in
the sketch.

4.2.1   Create Panel

The options of the CREATE strip (Fig. 4.5) allow the addition of new elements to the
canvas of the 2D sketch that is being designed. To make use of such tools, it is only
required to click on such options and later make a set of clicks on the main canvas.

Depending on the selected option, a sequence of clicks is required to define the properties of the shape designed. Additionally, it is possible to add measures or restrictions

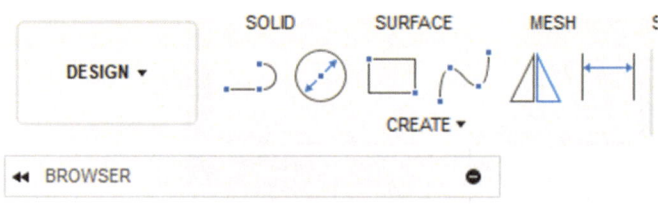

Fig. 4.5 Panel CREATE on the contextual strip SKETCH

while introducing the sketch to the canvas, reducing the time required to define all the elements. Nevertheless, such restrictions could be added later after an initial introduction of the elements to the canvas is made.

Finally, it is worth mentioning that, depending on the software, it is common that after hovering the mouse pointer over the options, a pop-up window appears, providing additional information on the objective of each operation, thus simplifying its memorization.

4.2.1.1 Line Creation

There is a set of widely used primitive shapes, among them, lines, circles, and rectangles are usually the most popular. When a line is required, it is only necessary to select the line option (Fig. 4.6) and then click twice, once where it is necessary to start the line and once at the end of it.

In addition, as can be seen in the help box of the line option, the line requires several constraints to be defined entirely. These restrictions can be the start and/or endpoint positions, the inclination in degrees concerning other geometry, the length of the line, or even some constraints such as parallelism, perpendicularity, horizontality, or verticality of the line.

Additionally, it is convenient to note that in the case of Fusion 360, the line operation allows the creation of multiple connected lines by employing a set of clicks until the "Esc" key is pressed, allowing in such a way a faster definition of the shape.

4.2.1.2 Circle Creation

Similar to how a line is created, to create a circle, it is only necessary to click on the circle option (Fig. 4.7) and then click on the main canvas. By default, the "center and radius" option will be selected, since this option is usually the most common. As its name indicates, this type of option will allow the addition of the circle by defining the center point on the first click and then defining the radius with the second one. Additionally, it is common for CAD software to request or indicate a provisional measurement of the radius or diameter of the circle so that the characteristics of the geometric primitive can be defined/restricted at the time of creation.

Additionally, constraining a 2D sketch thoroughly is highly recommended, as this ensures that the drawn elements will be precisely in the position that was indicated and, therefore, that it will not let any degree of freedom in the design (which could cause some undesired changes on the design during the manipulation of the sketch).

Finally, in order to constrain a circle, the "tangent" or "coincident" constraint is usually employed, as these constraints easily define the exact position of the perimeter or center of the designed circle.

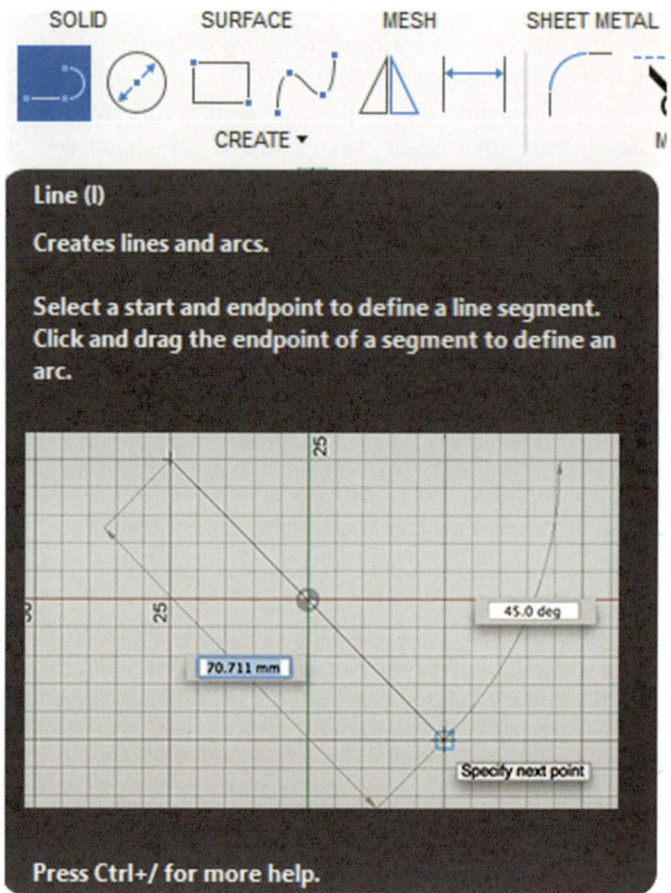

Fig. 4.6 Option line in the panel CREATE

4.2.1.3 Rectangle Creation

To create a rectangle, it is required to select the "Rectangle" option (Fig. 4.8) and then click on the design canvas. By default, the "rectangle by two vertices" option will be selected. However, the "center and vertex" option is also available if required.

It is important to note that traditionally, the design software automatically draws rectangles with their sides arranged horizontally or vertically. Thus, if an angled rectangle is required, using a set of four lines might be more convenient, with one at the desired inclination and the other with perpendicular constraints.

4.2.1.4 Fit Point Spline Creation

To create a fit point spline, one must select the "Fit Point Spline" option (Fig. 4.9) and then click on the points where the curve is required to pass through. After that, one only

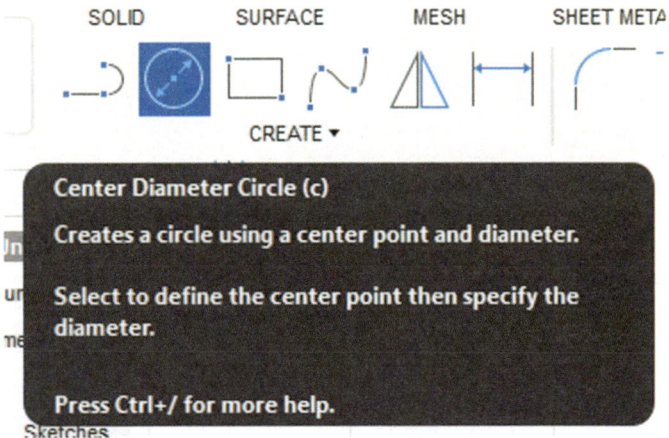

Fig. 4.7 Circle option on the CREATE panel

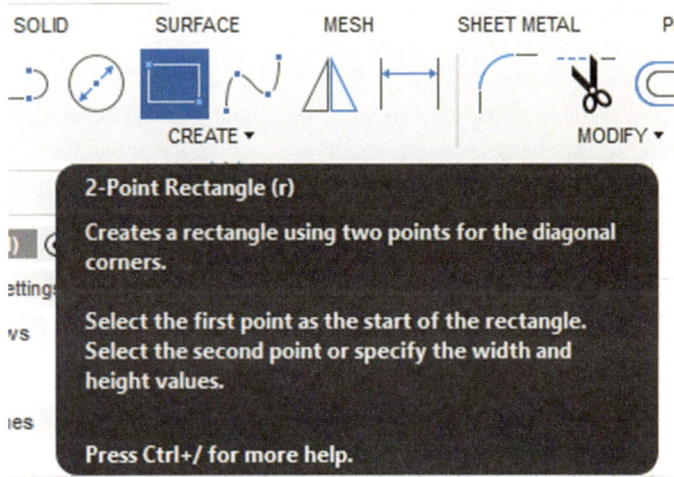

Fig. 4.8 Rectangle option in the CREATE panel

needs to click "Enter" to end the curve. Once the curve is positioned, it is possible to fine-tune the position of the points utilizing the point indicators or adjust the curvature of the line by using the green handles that appear around each point.

It should be noted that if one wants to print an object with curves, the printer's ability to follow these shapes is influenced by the orientation of the part, e.g., curves that are generated in a plane parallel to the printing surface (XY plane) have inherently good results with good shape following. However, they are affected by lateral expansion. On

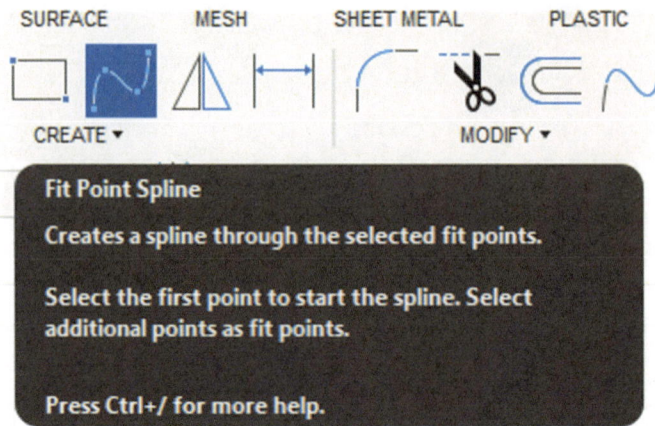

Fig. 4.9 Fit point Spline option in the CREATE panel

the other hand, curves generated using multiple layers on the Z axis tend to generate "steps," which can be inconvenient in some applications.

To illustrate this behavior, it is convenient to consider the 3D model presented in Fig. 4.10a. With it, a comparative analysis could be made by observing how the orientation present in Fig. 4.10b has better printing results concerning the orientation of Fig. 4.10c. This improvement can be explained by observing that in the first orientation, the part has a curve in concordance with the XY plane, on which the printer can provide better tracking. In contrast, in the second orientation, the curve must be expressed using multiple printing layers, which implies a lower resolution and, therefore, a worse result.

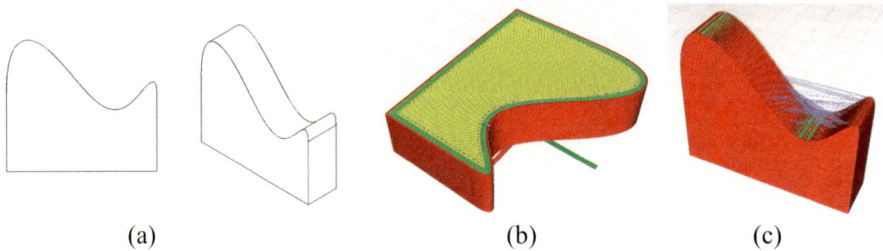

(a)   (b)   (c)

Fig. 4.10 Comparative example of the implications of the printing orientation of a model concerning the 3D printer's ability to express curved objects (a) Model to be printed (b) Orientation that offers better tracking of the curves (c) Orientation of worst resolution for curves

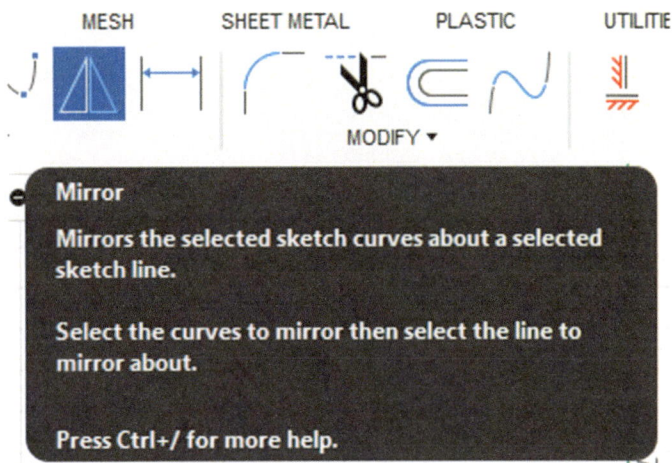

Fig. 4.11   Mirror option in the CREATE panel

4.2.1.5   Mirror

In many designs, it is usual to find symmetry patterns. Based on this standard feature, most CAD design software offers a tool to define these structures, reducing the overall design time. Thus, the Mirror option (Fig. 4.11) allows the generation of symmetrical structures concerning a reference shape and a symmetry axis.

To make use of this option, so-called "construction" lines are usually employed. These types of lines allow the user to define geometries that, although they facilitate the establishment of relationships between bodies in the 3D model, do not form a physical part of it.

Switching between a construction line and a structural line is as simple as selecting the line or curve of interest and pressing "x," which visually changes the curve from continuous to dashed, indicating that it is a construction line.

An example of the use of the Mirror operator and construction lines can be observed in Fig. 4.12. There, the body to the left of the construction line is taken as the reference shape for the reflection, and the construction line is taken as the axis of the reflection. After applying this operation, it is possible to observe the new shape with the desired symmetric properties.

4.2.1.6   Sketch Dimensions

Since CAD design is oriented to generate parts that will be built in the physical world, it is necessary to establish measurements on the parts and not only the proportions between them. With this in mind, the "sketch dimensions" element (Fig. 4.13) allows

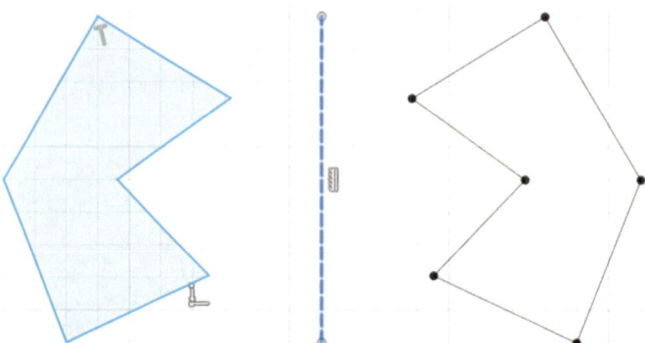

Fig. 4.12 Example of the use of the operation Mirror over an irregular polygon

the generation of various measurement constraints, which range from longitudinal distances (straight lines between two points or distances on some axis of interest) to the definition of diameters or radii in arcs.

To switch between the distinct types of constraints, one can move the cursor to adjust for the measurement on a specific axis of interest or seek to define the length of a drawn line. In the case of arcs or diameters, the dimension type will dynamically adjust to radius or diameter depending on the primitive type.

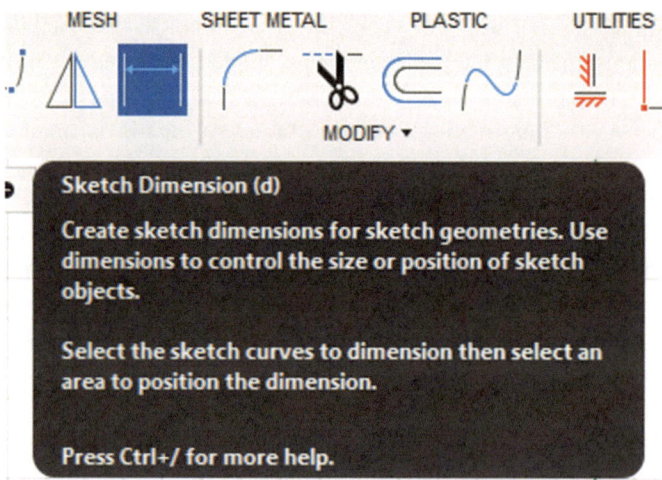

Fig. 4.13 Sketch dimension option in the CREATE panel

4.2.1.7 Additional Options of the Create Panel

Finally, if a more complex shape is required, it is possible to go to the different additional options in the CREATE section by clicking on the text "CREATE" (Fig. 4.14). Among these additional options, it is possible to see tools to generate arcs, inscribed or circumscribed polygons, ellipses, and texts, among other tools. Additionally, there is also a subsection to generate rectangular patterns (as in a grid) or circular patterns (like petals on a flower).

4.2.2 Operations of the Modify Panel

Once the primitive shapes that compose the 2D sketch have been defined, it may be necessary to apply a modifier to adapt the sketch to the needs of the model, e.g., by rounding a vertex of the model, by connecting two edges through a contour curve, or by trimming a line. The "modify" panel (Fig. 4.15) offers several solutions for all these operations.

Fig. 4.14 Additional Options of the CREATE strip

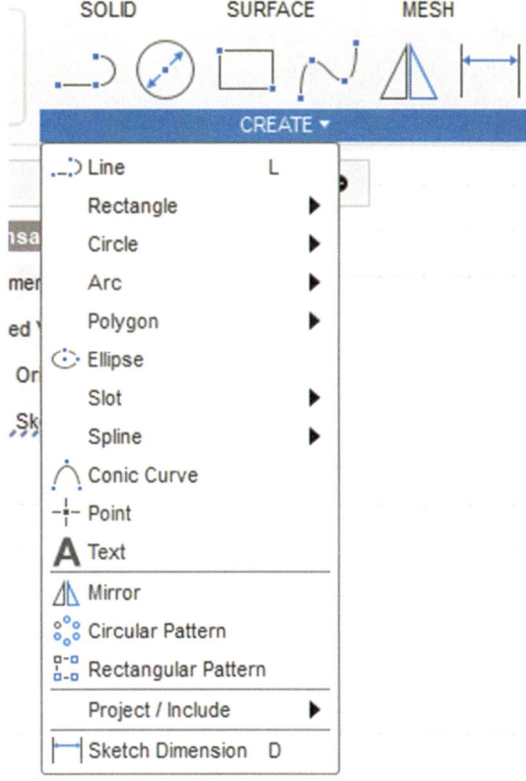

Fig. 4.15 Panel MODIFY on
the contextual strip SKETCH

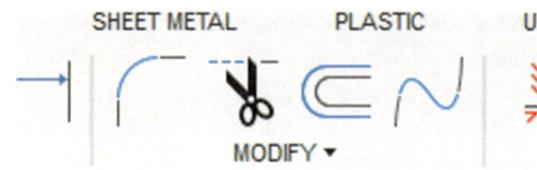

4.2.2.1 Fillet Modifier

A frequently required operation is the Rounding of a 2D sketch vertex. To use this modifier, one can click on the fillet modifier (Fig. 4.16) and then, with the modifier active, select the two edges of the vertex to be rounded.

It is important to remember that, in case two intersecting lines generate the vertex and not just a corner, the edges to be preserved after the operation are those selected by the click, cutting the prolongations of the lines in the process. Thus, in Fig. 4.17, it can be seen how by having an intersection of two lines (Fig. 4.17a) and selecting the upper and left extensions concerning the crossing (circles on Fig. 4.17b), a rounded vertex (Fig. 4.17c) is obtained, which can be modulated (using a numerical parameter or the available handler) so that the Rounding has the radius of interest.

Thus, this 2D modifier or its 3D equivalent tends to be widely implemented, as it ensures that the parts do not have sharp edges that can hurt people while the object is being manipulated. However, in the case of applying it to a model that will be 3D printed, it is convenient to consider that depending on the orientation of the part and the radius used, the part may require a post-processing process to solve the problems generated by the low printing resolution between layers.

Finally, it is also essential to consider that the use of this modifier is not recommended when it involves generating surfaces detached from the printing bed (e.g., a cube rounded in all its vertices and edges Fig. 4.18), since due to thermal issues and the support required in that region, some deformation may be generated on the surface of the part.

Fig. 4.16 Fillet modifier in
the MODIFY panel

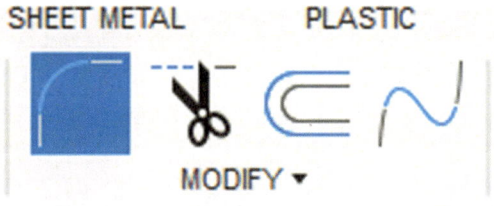

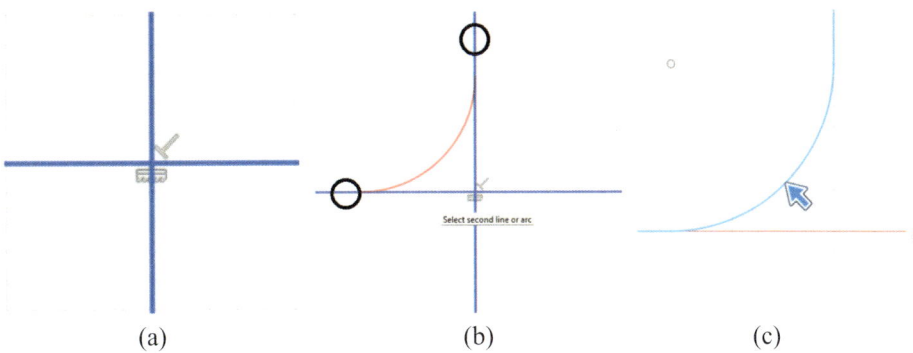

(a)　　　　　　　　(b)　　　　　　　　(c)

Fig. 4.17 Example of the use of the Fillet modifier on a couple of intersecting lines (a) Original state (b) Selection of the upper and left segments (c) Result of the modifier

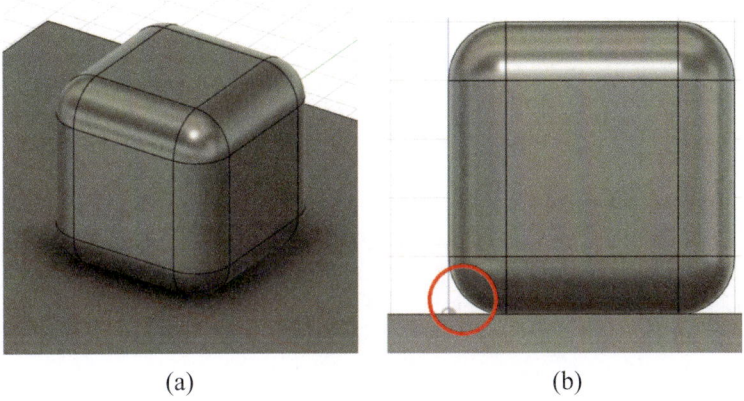

(a)　　　　　　　　　　　　(b)

Fig. 4.18 Example of a model where it would be desirable to avoid roundings (a) Isometric view (b) Lateral view with a marker on the region where support may be required

4.2.2.2 Trim Modifier

When the primitives have already been arranged in the sketch, it is usual to require cutting segments of a primitive to generate, for example, an arc from a circle, or to remove an excess of a line after applying a tangency constraint to a circle. Thus, using the Trim operation is convenient for those scenarios (Fig. 4.19).

To cut the excess segment using this modifier, one simply clicks on the Trim button and then locates the cursor on the segment to be removed, which will be highlighted as shown in the comparative view in Fig. 4.20.

Fig. 4.19 Trim modifier in the MODIFY panel

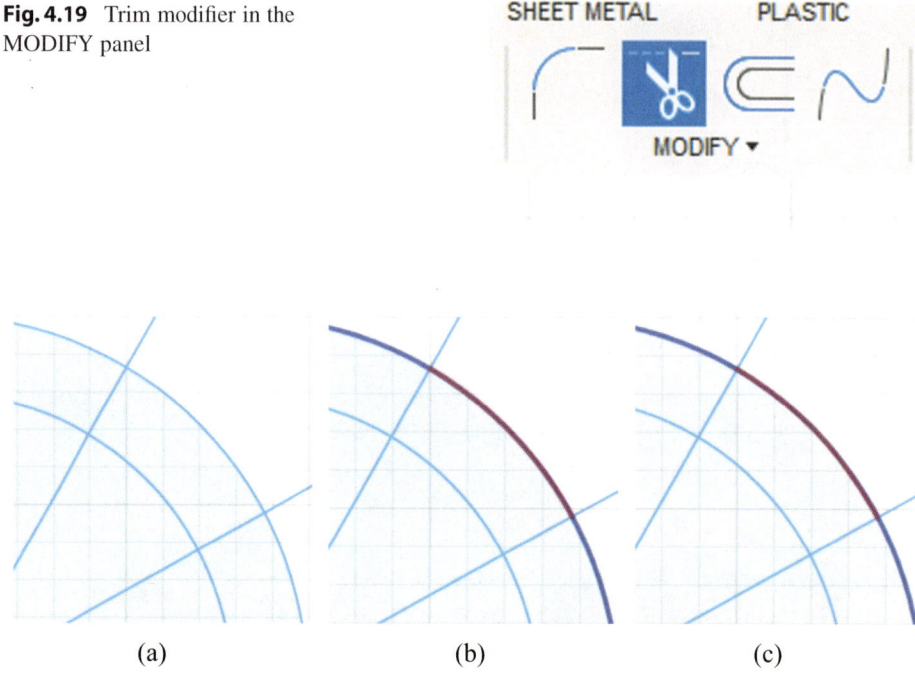

(a)   (b)   (c)

Fig. 4.20 Example of the use of the Trim modifier (a) Original State (b) Selection of the element to cut (c) Result of the application of the modifier

4.2.2.3 Offset Modifier

Considering physical objects cannot be infinitesimally thin, it is common to set thicknesses in sketches. Thus, the offset modifier (Fig. 4.21) allows the generation of adaptive contours that take the desired thickness as a parameter and generate a new contour. Additionally, this parameter accepts positive or negative values depending on whether the new perimeter is oriented inside or outside the base figure, easing its use in such a way.

Thus, Fig. 4.22 shows an example of the use of the Offset modifier for positive and negative values. There, it can be seen how generating edges from a base geometry is possible. Additionally, it does not matter if the shape is regular or irregular (like the one presented). Finally, it is also convenient to point out that the offset modifier does not

Fig. 4.21 Offset modifier in the MODIFY panel

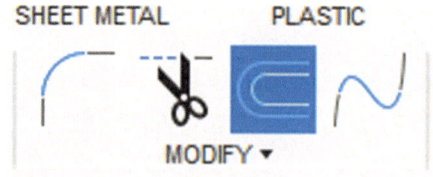

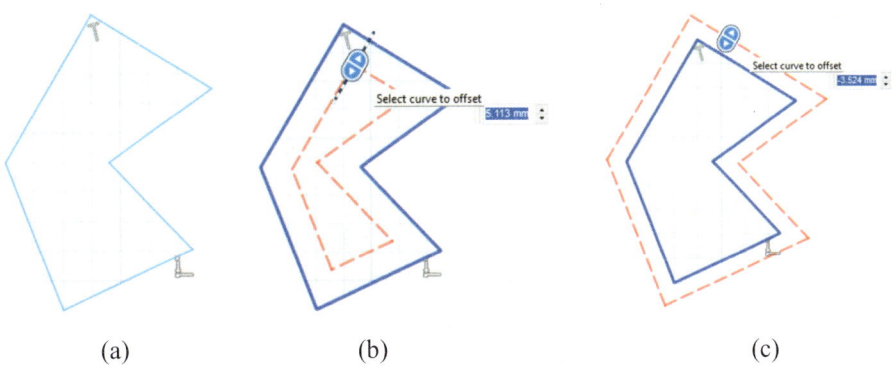

(a)   (b)   (c)

Fig. 4.22 Example of the use of the Offset modifier (a) Original state (b) Example of an inward offset (c) Example of an outward offset

Fig. 4.23 Bend modifier in
the MODIFY panel

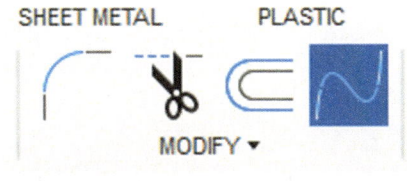

require the figure to be a closed geometric body so that it can operate even on open curves.

4.2.2.4 Blend Curve Modifier

On the other hand, it is also common to find scenarios where one wants to connect two segments of primitives employing a curve. Although this problem can be solved with a Fit Point Spline, the use of the Blend curve modifier (Fig. 4.23) allows the definition of these curves in a faster and more efficient way; since it considers that the limits of these curves must be tangent to the indicated edge (generating continuous lines that avoid abrupt edges).

An example of using this modifier can be seen in Fig. 4.24, where it can be seen how easy it is to define this curve, simplifying the design process and reducing the time required to indicate the properties of the sketch.

4.2.2.5 Additional Options of the Modify Panel

Finally, if other modifiers are required, it is possible to click on "MODIFY" (Fig. 4.25) within the modify panel to obtain a drop-down menu that offers modifiers to facilitate the manipulation of primitives, allowing for example, the extension of primitives, the drawing of sketches, or even the definition of parameters to generate modifiable models on demand.

Fig. 4.24 Example of use of
the modifier Blend curve

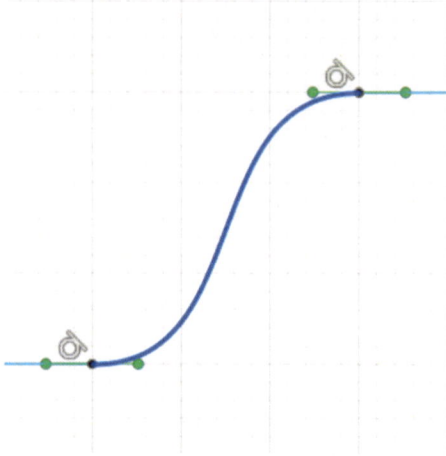

Fig. 4.25 Additional options
of the MODIFY panel

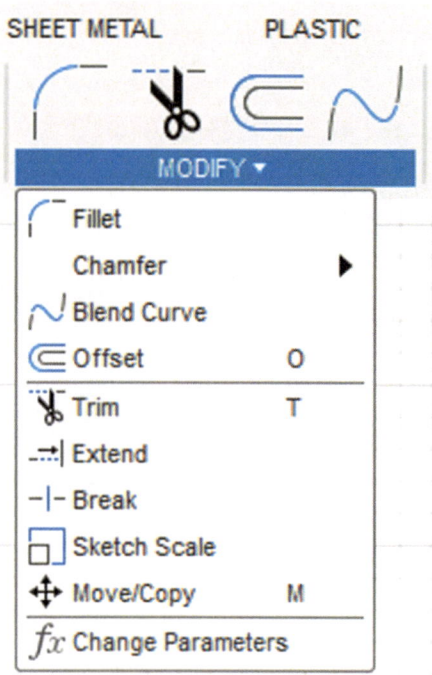

4.2.3   Operations of the Constrain Panel

In addition to the use of measurements, it is also possible to restrict the properties of a sketch by means of geometric constraints. In the case of Fusion 360, these constraints are available in the "CONSTRAINTS" panel (Fig. 4.26). The purpose of these constraints is to establish geometric relationships between the various components of the sketch, thus facilitating the process of defining how a feature relates to the rest of the object. Thus, avoiding any undesirably free degree overlooked would cause the resulting model to have some design errors or be under-restricted.

To facilitate early detection of designs that are not fully constrained, Fusion 360 offers a color change on sketch curves that are fully constrained. Thus, any element with its original color (light blue) will not be fully constrained. Additionally, it is usually convenient to click on the part and drag it with the mouse to facilitate the detection of the type of constraint to be used. This strategy will cause the model to move, rotate, or alter its dimensions, revealing the missing constraint.

To exemplify this property, Fig. 4.27a shows a body that initially appears to be wholly constrained. However, upon closer observation, it is possible to see that the right side of the rhombus is not defined. This lack of constraints can be corroborated by clicking on the light blue side and moving it, allowing us to observe that the x-axis of this feature is not defined, as shown in Fig. 4.27b. From this observation, it is possible to constrain the lower right corner of the rhombus to be coincident with the corner of the rectangle, thus obtaining Fig. 4.27c. Subsequently, it can be observed that there was no parallelism relation between the angled edges. Thus, the entirely restricted body can be obtained by adding this restriction, as shown in Fig. 4.27d.

4.2.3.1   Horizontal/Vertical Constrain

The Horizontal/Vertical constraint (Fig. 4.28) allows one to specify to the CAD drawing software that the selected line should be parallel or perpendicular to the horizontal of the sketch.

It is important to note that the horizontality or verticality of the line is selected considering the current state of the line to which the constraint is applied, e.g., if the line is in a position similar to a horizontal line, then the software determines that a horizontality constraint is desired. However, if the line was initially in a position similar to a vertical line, it determines that a verticality constraint is desired.

Fig. 4.26 CONSTRAINTS panel on the contextual strip SKETCH

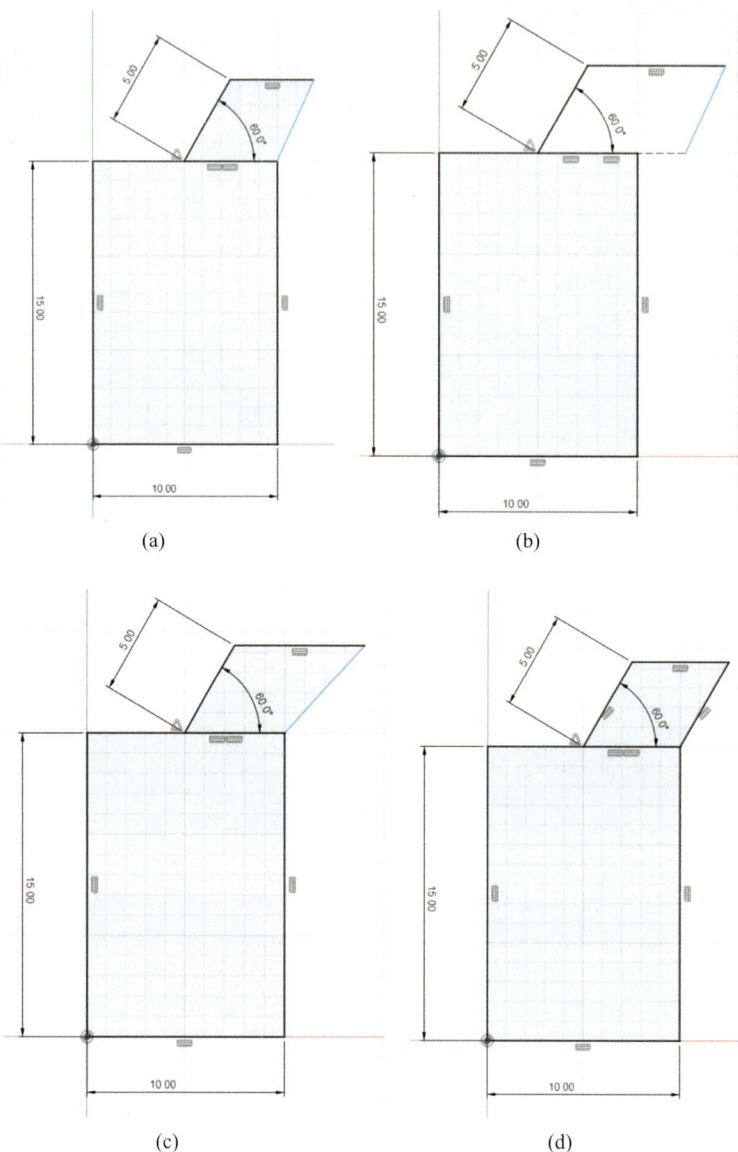

(a)   (b)

(c)   (d)

Fig. 4.27 Example of the use of restrictions to completely define a sketch

Fig. 4.28 Horizontal/Vertical constrain in the CONSTRAIN panel

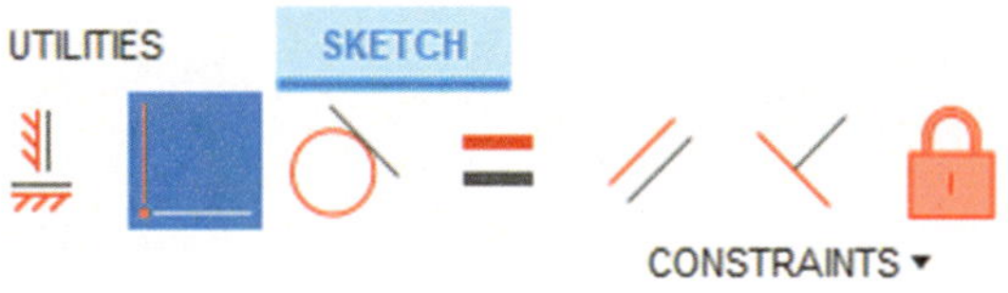

Fig. 4.29 Coincident constraint in the CONSTRAIN panel

4.2.3.2 Coincident Constrain

The coincident constraint (Fig. 4.29) is one of the most used constraints, because it allows the designer to specify to the CAD design software that two elements are located or pass through the same position, which is a common requirement. Thus, to use this constraint, one selects it from the constraints panel and then clicks on the elements to be constrained. By default, if possible, Fusion 360 takes the first object as the base and the second as the element to be adjusted to meet the constraint.

Although there are multiple scenarios where the use of this constraint facilitates the design, some of the most common scenarios consider the merging of vertices of two primitives (Fig. 4.30a and 4.30b) or defining some point that should intersect a line or curve (Fig. 4.30c and 4.30d).

4.2.3.3 Tangent Constrain

The Tangent constraint (Fig. 4.31) is a very versatile constraint when using circles in the 2D sketch, as this constraint allows the restriction of the shapes of the sketch to touch other curves but not cross them.

Although simple, this constraint is powerful for defining relationships between geometric bodies. An example of its use can be seen in Fig. 4.32a and 4.32b. These figures show a point of apparent tangency (Fig. 4.32a), which, after applying the tangent constraint (Fig. 4.32b), can generate relationships of complete restriction. Additionally, in Fig. 4.32c, it is easy to see the practical implications of this constraint and its multiple uses, ranging from simple roundings to the generation of complex curves continuously interconnected by tangent constraints.

4.2.3.4 Equal Constrain

The equal constraint (Fig. 4.33) allows the designer to specify to the CAD design software that one desires two different elements to have identical dimensions.

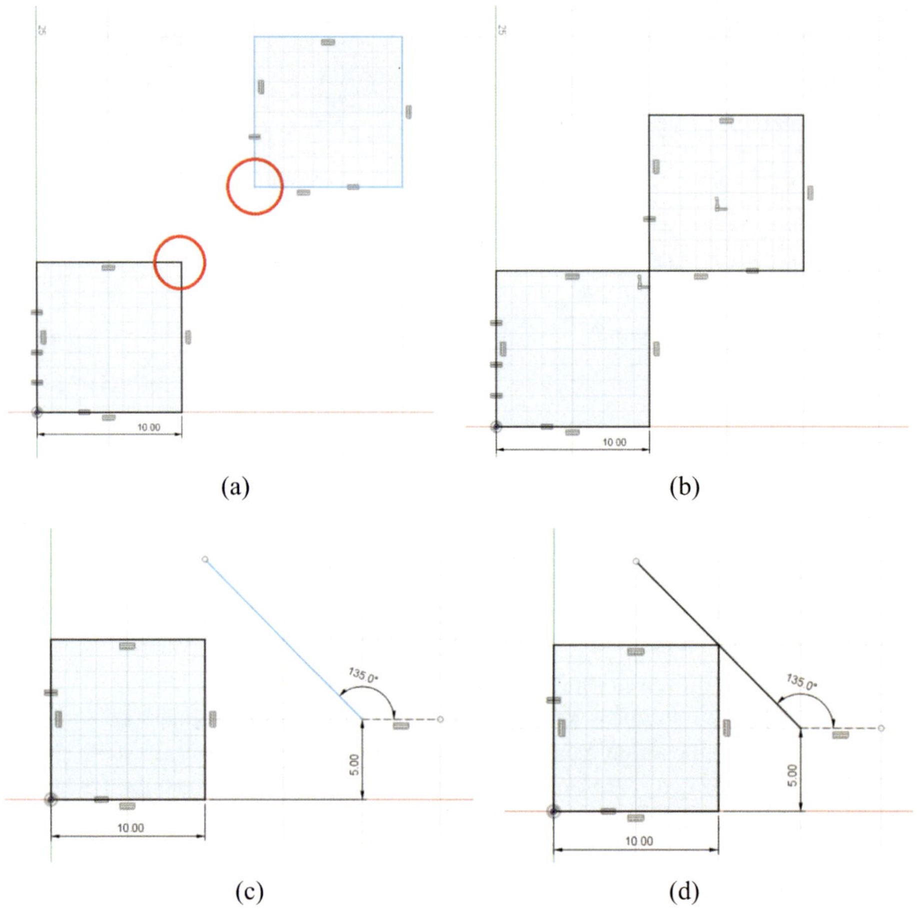

Fig. 4.30 Example of the use of the constrain coincident (a) Original state of the primitives with markers on the points selected for the constraint (b) Result of the merging of the points (c) State before the addition of the restriction (d) Result of making the line and que corner of the primitive coincident

Fig. 4.31 Tangent constraint in the CONSTRAIN panel

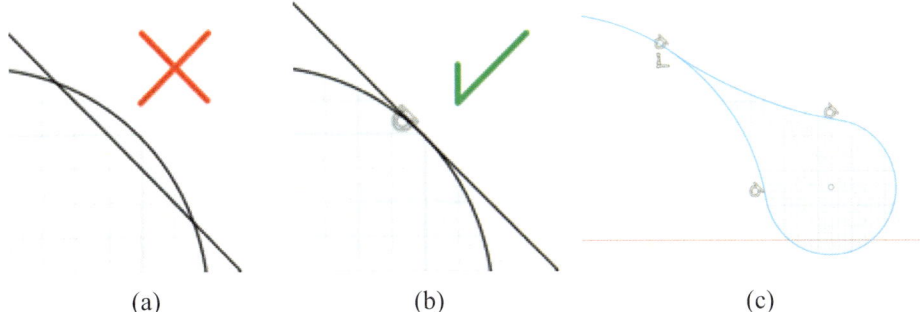

(a)   (b)   (c)

Fig. 4.32 Example of the use of the tangent constraint in different environments (a) Close-up of an example of false tangency (b) Close-up of an example of true tangency (c) Example of the use of tangent constraints in a design sketch

Fig. 4.33 Equal constraint in the CONSTRAIN panel

This element is helpful as it allows multiple model parts to be modified by changing a single reference. For example, it is very convenient to indicate that all bolt holes in a model have equal metrics since this would later allow the modification of all referenced elements by altering only one of them instead of modifying each dimension of each one individually.

4.2.3.5   Parallel and Perpendicular Constraints

Finally, the parallel (Fig. 4.34a) and perpendicular (Fig. 4.34b) constraints allow for establishing relationships between the slopes of any two lines. While the first constraint (parallel) forces the system to have two lines with an identical slope, the second constraint (perpendicular) tells the CAD system that one line must have the inverse slope of the other.

4.2.3.6   Other Constraints of the Constraint Panel

In addition to the constraints presented, there are other constraints that may have a greater or lesser use among designers but can promote fast development. These additional options may include midpoints, concentricity, collinearity, or other interest constraints. If these constraints are required, one can click on the CONSTRAINTS panel (Fig. 4.35) to display a drop-down menu with these additional options.

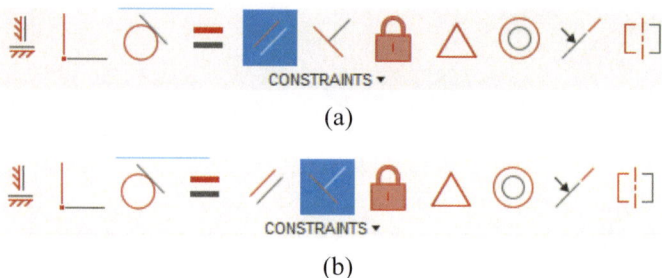

Fig. 4.34 Parallel and perpendicular constraints in the CONSTRAIN panel (a) Parallel (b) Perpendicular

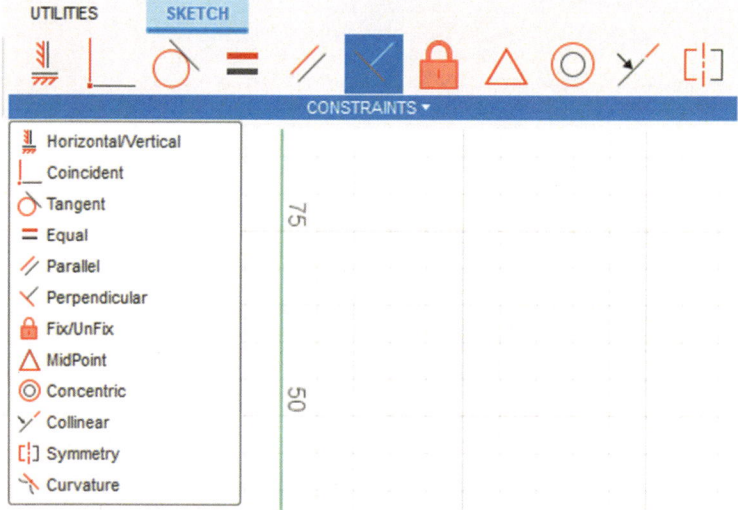

Fig. 4.35 Drop-down menu of the CONSTRAINTS panel with additional constraints

Finally, it is convenient to point out that most CAD software usually seeks to establish the constraints as the sketch is drawn on the canvas, thus facilitating the design process and reducing the number of degrees of freedom as they are added.

4.3   2D Sketching Exercises

Since the mastery of 2D design is strongly related to the exercise of such activity, three 2D sketch designs are presented below. Thus, it allows the exercise of various skills related to design, which are recommended to be completed to improve the perception and capabilities of 2D CAD drawing before trying to 3D model an object.

4.3.1   Mosaic Star by Mirror

In this exercise, one must generate a mosaic star using the mirror option. To do this, initially, one needs to make a trapezium like the one shown in Fig. 4.36a, which uses four lines with the base and height sizes indicated. Additionally, one must indicate that the major and minor bases are horizontal, and the sides are perpendicular, measuring the same length.

Once one has the trapezium, one can select the Mirror option from the CREATE panel to generate a second trapezium concerning the major base, as shown in Fig. 4.36b. It is important to note that only the minor base and the two sides should be selected for mirroring; otherwise, the line of the major base would be duplicated.

Subsequently, it is possible to repeat this operation, but considering the left side of the lower trapezium as the reflex line, the result of such operation is presented in Fig. 4.36c.

Finally, it is possible to repeat this procedure considering as the line of reflection one of the two sides that form the diagonal that crosses the origin, thus obtaining the mosaic star shown in Fig. 4.36d.

4.3.2   Camera Diaphragm

In this exercise, one uses n-sided polygons, circular patterns, circles, and arcs. Initially, a hexagon inscribed in a 50 mm circle must be generated using the "polygon" tool of the additional options of the CREATE panel (Fig. 4.37a). It is important to note that this hexagon must be oriented vertically (its sides) utilizing the "Horizontal/Vertical" constraint.

Subsequently, it is necessary to create a circle that encompasses the hexagon, having a diameter of 100 mm, as shown in Fig. 4.37b.

Once this base shape is available, an arc should be created with its center on the lower part of the outer circle (one may require to generate a vertical construction line as a reference), touching with its perimeter the opposite side of the hexagon and the perimeter of the outer circle, as shown in Fig. 4.37c.

Finally, it is possible to generate a circular pattern of six repetitions by selecting the generated arc and the center of the sketch, obtaining the result shown in Fig. 4.37d.

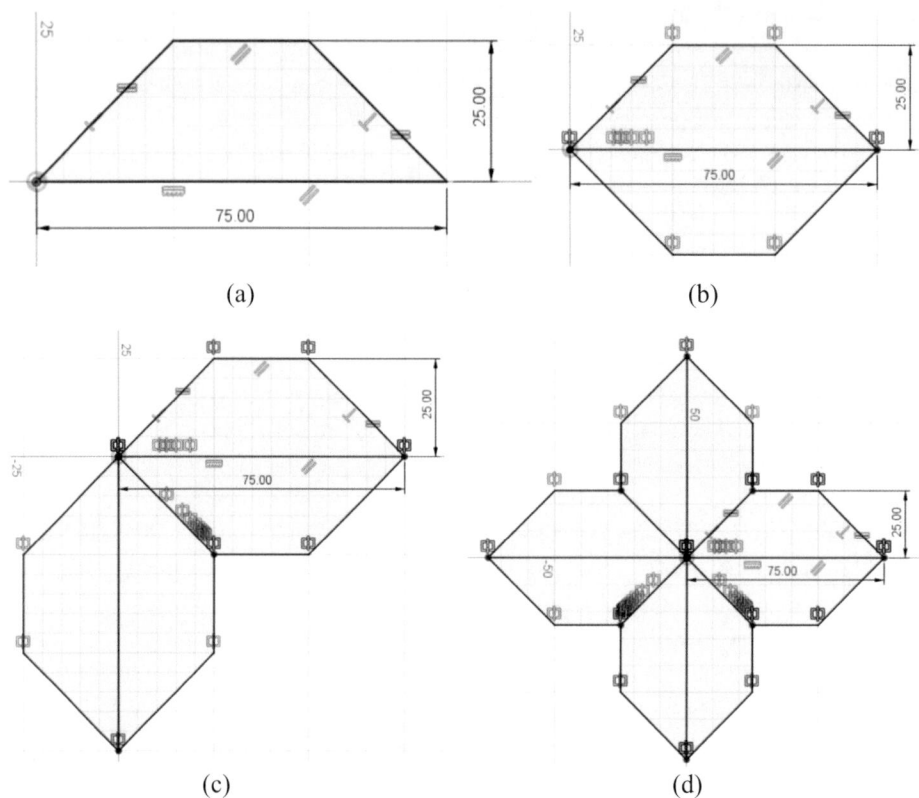

Fig. 4.36   2D Sketching exercise 1

4.3.3   Fractal Vise Gripper

In this exercise, one makes use of circles, arcs, offset modifiers, and tangency constraints. Initially, two circles of 20 mm and 10 mm in diameter must be generated. After that, it is required to align them with a horizontal line and separate them with a distance between centers of 40 mm, as shown in Fig. 4.38a.

After that, an offset modifier should generate an outline for each circle, considering a spacing of 5 mm, as shown in Fig. 4.38b. Once these base elements are available, it is necessary to generate an arc that is tangent to both circles and has a center coincident with the horizontal construction line (Fig. 4.38c).

Finally, a last arc is generated between the circles, indicating a tangency between them and a radius of 15 mm, as shown in Fig. 4.38d.

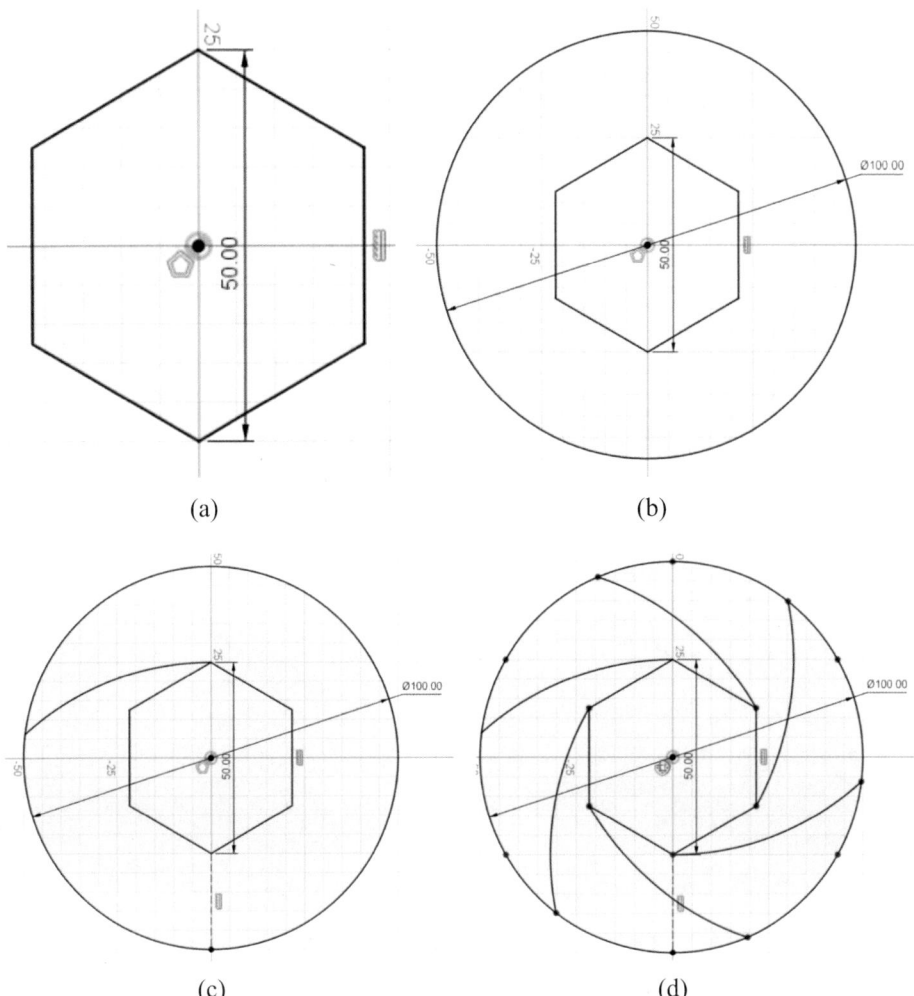

Fig. 4.37   2D Sketching exercise 2

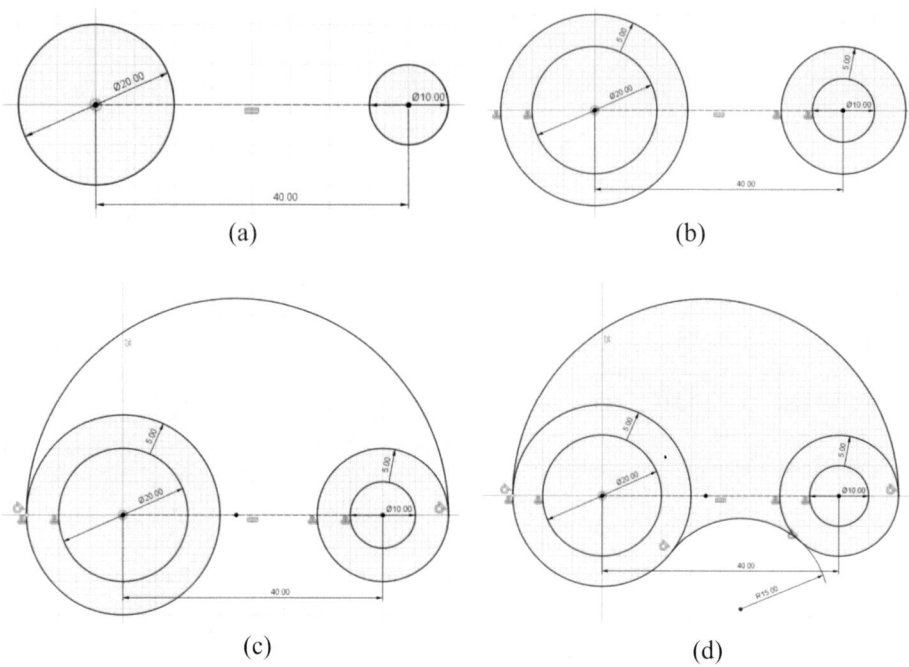

(a)   (b)

(c)   (d)

Fig. 4.38   2D Sketching exercise 3

References

1. L. Olsen, F.F. Samavati, M.C. Sousa, J.A. Jorge sketch-based modeling: a survey. Comput. Graph (Pergamon), **33**(1), 85–103 (2009). https://doi.org/10.1016/j.cag.2008.09.013
2. Fusion 360—autodesk, Set up a static stress analysis [Online]. Available: https://help.autodesk.com/view/fusion360/ENU/?guid=SIM-SSA. Accessed May 15, 2024
3. M. Zawada-Michałowska, J. Kuczmaszewski, S. Legutko, P. Pieśko, Techniques for thin-walled element milling with respect to minimising post-machining deformations. Materials. **13**(21), 1–17 (2020). https://doi.org/10.3390/ma13214723
4. T.D. Tang, Algorithms for collision detection and avoidance for five-axis NC machining: a state of the art review. CAD Comput. Aided Des. **51**, 1–17 (2014). https://doi.org/10.1016/j.cad.2014.02.001

3D Design in CAD Software for 3D Printing

5

5.1 Considerations and Skills for 3D Design

Once one has a solid foundation in the design of 2D sketches, it is possible to begin the journey into generating 3D structures that can accurately reflect the features of the models to be designed, considering a set of constraints, measurements, and physical properties of the object. For this chapter, such a 3D journey employs the tools available on the CAD software from Autodesk "Fusion 360".

While 2D sketching tends to be a flexible process by nature, 3D design requires special attention from the designer to consider at least the following two practical elements: the malleability of designs and the physical properties intrinsic to the shape and weight of the model.

The first element, malleability, consists of generating designs that can quickly adapt their features. The capacity to make malleable models is a valuable skill, as it is expected to make multiple prototype versions before settling on a definitive version that passes all the required tests [1]. Thus, taking care to generate constraints and operations that allow features to be easily modified impacts the time required to complete a project. This detail is of paramount importance when the parts are 3D printed, as the filament can generate unwanted expansions or contractions [2], which can directly affect the fit between different components and influence their behavior, especially on moving components. Considering the above, malleability is essential to solve these inconveniences in prototype versions.

On the other hand, it is also important to consider the physical properties of the element under construction. This consideration could simplify component manufacturing by, for example, joining or splitting components to simplify their construction. An illustrative analysis of such consideration is presented in Fig. 5.1, where two models can offer a solution for a problem. However, the first employs only one component, and the second

E. Cuevas et al., *DC Motors*, Synthesis Lectures on Engineering, Science, and Technology, https://doi.org/10.1007/978-3-031-64354-5_5

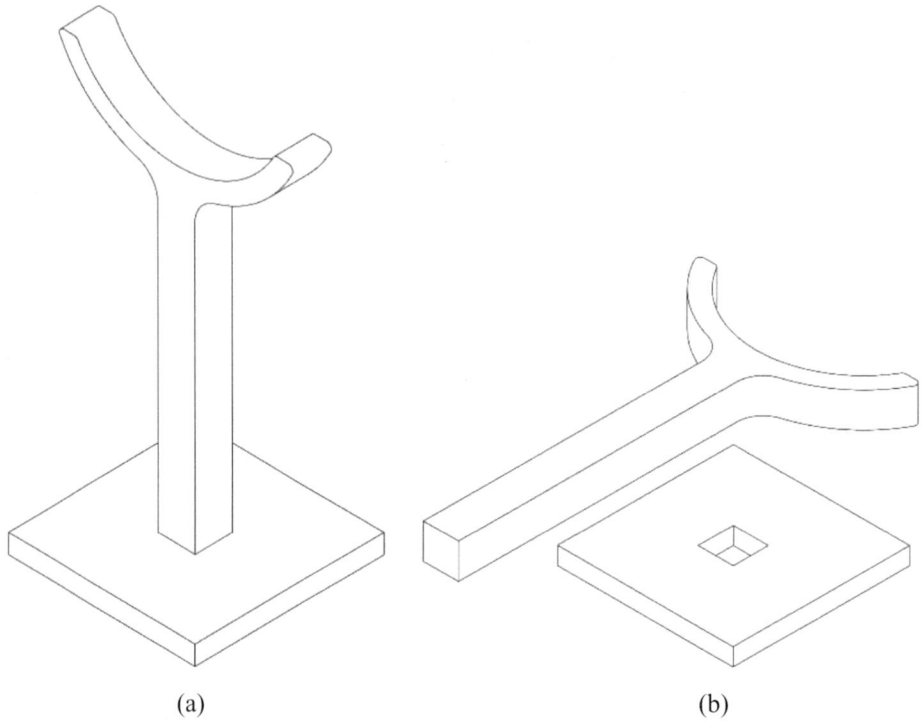

(a)   (b)

Fig. 5.1 Comparison between two ways to build an object. **a** As one component. **b** As two split components

considers two. In the case of Fig. 5.1a, the part is designed as a single component, implying it must be 3D printed in a single operation. On the other hand, Fig. 5.1b is designed to print the element in two components, which must be joined after the 3D printing process.

While the first alternative offers the advantage of not requiring additional processes and that the model is composed of a single piece, this also implies the disadvantage of requiring a large number of supports that serve as "scaffolding" for all those elements that do not have material under them (in this case, the upper curve of the model). Although there are techniques to reduce their impact [3], these supports imply higher material consumption and leave a mark on the 3D object once removed.

Moreover, printing tall objects is an operation that should be avoided as much as possible in 3D printing [4]. This is due to the fact that the taller an object is, the more likely it is that the lateral forces generated by the depositing material during printing will cause the print to peel off the printer bed [4]. Additionally, the taller a model is, the more layers it will require, which allows errors from lower layers to accumulate and cause some unwanted noticeable warping in flat regions of the printed model.

On the other hand, the second alternative (Fig. 5.1b) offers the option of parallel printing of the two components. Thus, although this option has the disadvantage of requiring additional processes to adhere the components, it also has the advantages of not requiring supports, implying a low height printing in the models, providing a large contact area with the printer bed (which improves the adhesion of the model to the bed) and offering an orientation of layers that increases the resistance to fractures in the model.

The last observation has major implications on the performance of the 3D printed model, as printed objects have different resistance properties depending on how forces are applied and how the object printing was oriented [5], affecting its strength positively or negatively.

A practical example of this property can be observed when cutting meat or some vegetables with a knife. When the knife is oriented along the fibers, it cuts effortlessly; however, when the cut is made perpendicular to the fibers, it requires a high amount of force and does not generate a clean cut. Similarly, when the model is printed in a particular arrangement, a plane is generated in which the print is strong, while the remaining axis offers less resistance to stress.

Thus, for the example case presented, printing with the layer orientation as in Fig. 5.1b offers a better response to lateral loads, since its printing orientation allows the layers to be aligned conveniently to increase its resistance compared to the orientation of Fig. 5.1a.

Given the above, it can be seen that 3D design is a complex task requiring the designer to acknowledge various printed-related phenomena. Several of these phenomena will be discussed during this and subsequent chapters to provide intuitive perspectives on them, using a handful of model examples for such an objective.

On the other hand, this chapter will focus on creating and modifying 3D objects from 2D sketches, giving examples of real-world objects with such shapes or modifiers.

5.2   Desing Solid 3D Objects and Assembling Them

Similar to the contextual menus of 2D sketch design, it is possible to find a contextual ribbon for solid body design (Fig. 5.2). This ribbon contains multiple options specialized in different tasks or operations. However, in a first approach to 3D design, and considering the focus of this book, it is only necessary to have a deep knowledge of some of these options, while the rest can be reserved for more complex operations.

Fig. 5.2 Contextual ribbon with options for creating, modifying, and assembling different 3D components

5.2.1   Create Panel

The CREATE section (Fig. 5.3) contains most of the operations used to give volume to a design, or to replicate some feature of interest employing a pattern.

Although this section has multiple tools, the most used are extrusion and profile revolution.

5.2.1.1   Extrude

The extrude tool (Fig. 5.4) is one of the most used tools in 3D body design. This tool allows a 2D surface to be converted into a 3D feature by adding material on the surface until a desired height is reached, as shown in Fig. 5.5b.

On the other hand, this tool can also be used as a subtractive tool, removing material from a pre-existing body, as shown in Fig. 5.5c. In this mode, the generated body interacts with all the surfaces it encounters in its path, applying a subtraction to all of them in the shared volumes.

To use this tool, it is possible to employ the dynamic manipulators that appear after selecting the tool and the sketch on which it is to be applied. On the other hand, like many 3D operations, a pop-up window is expected to appear on the right side of the screen (Fig. 5.6), allowing the modification of the parameters of the applied operation numerically. Additionally, some 3D transformations usually offer variations of them on such a window, making them more suitable for some applications. For example, in the case of the extrude operation, this tool also offers options to generate extrusions in one or both directions concerning the reference plane, and allow these extrusions to be symmetrical or asymmetrical.

Finally, thanks to such a pop-up window, it is also possible to generate extrusions with some positive or negative angle at their edges, expanding or contracting the base surface according to the magnitude and direction of the requested angle. Thus, this capability

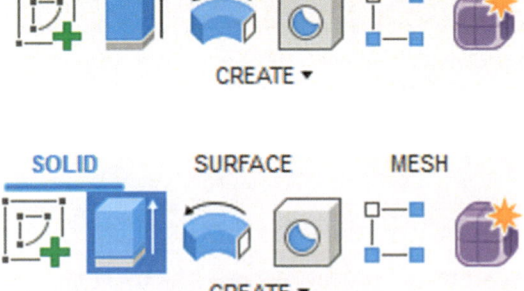

Fig. 5.3 CREATE panel on the contextual strip SOLID

Fig. 5.4 Extrude tool in the CREATE panel

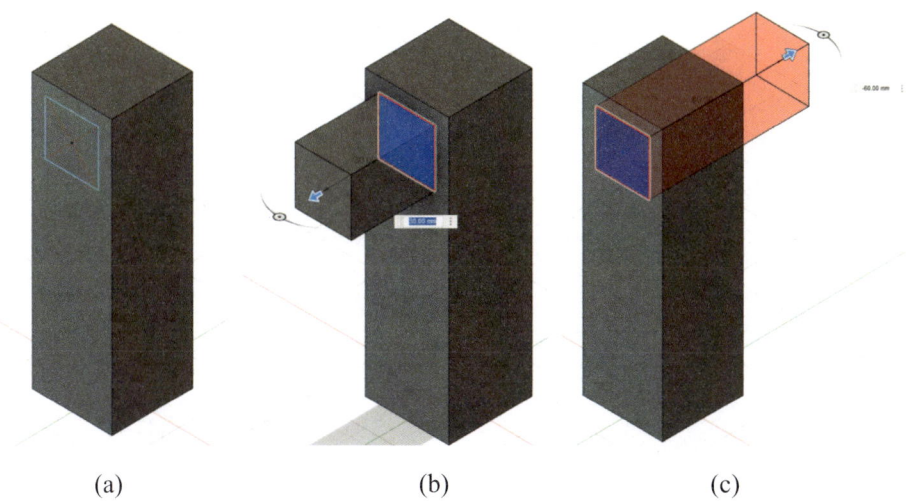

(a)   (b)   (c)

Fig. 5.5 Example of using the extrude tool to add or subtract volume. **a** Reference 2D sketch for the operation. **b** Extrude operation is used to add volume. **c** Extrude operation is used to subtract volume

Fig. 5.6 Pop-up window with additional options for the extrude operation

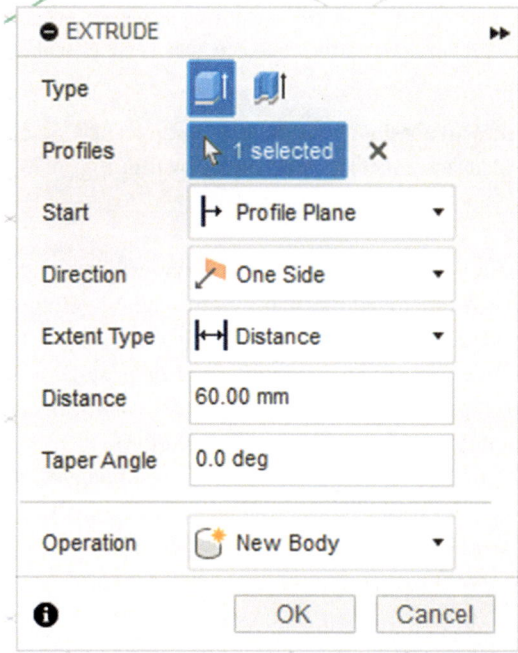

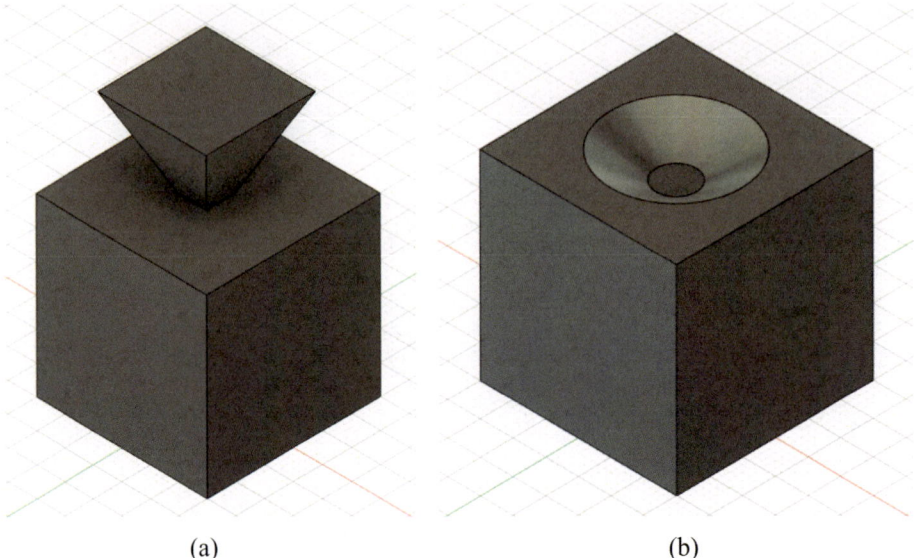

(a)   (b)

Fig. 5.7 Example of the use of positive and negative extrusion angles. **a** Use of positive angles in conjunction with an additive extrusion to generate a handle. **b** Use of a negative angle in conjunction with a subtractive extrusion to generate a countersunk hole

makes it possible to generate pyramid-like or cone-like shapes with simple square or circular reference sketches, as shown in Fig. 5.7.

5.2.1.2   Revolve

The revolve tool (Fig. 5.8) is in conjunction with the extrude tool, one of the most frequently employed mechanisms for adding volume to a 2D sketch. This importance is due to these tools referring to a widely used conventional manufacturing machine. While the extrude tool allows the building of parts suitable for milling machines (subtractive extrusion) or for 3D printers (additive extrusion), the revolve tool allows the building of parts suitable for lathes (bodies of revolution).

In a lathe, a base material is arranged on a support that rotates on an axis. This rotation allows the base material to interact with some wear element that removes its material until a desired shape is achieved.

Fig. 5.8 Revolve tool in the
CREATE panel

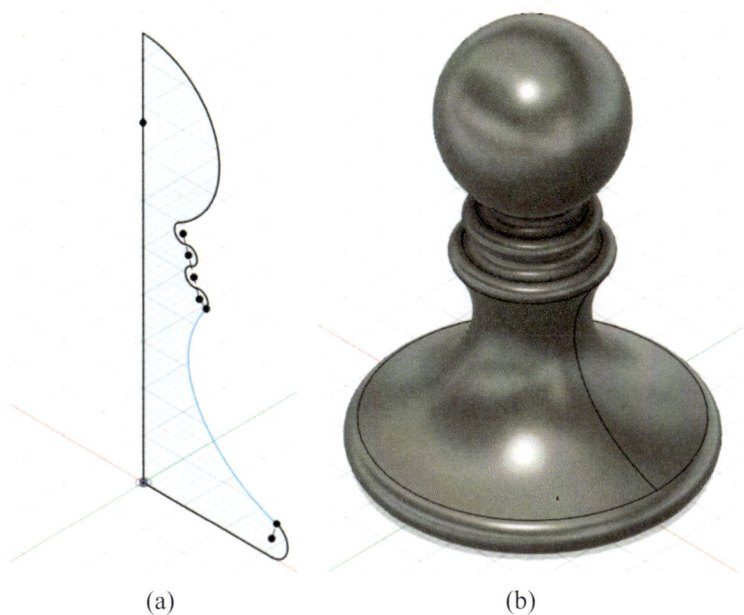

(a)   (b)

Fig. 5.9 Example of the use of the revolve tool. **a** 2D reference profile and axis of revolution. **b** Result of the application of the revolve tool

In order to use this tool, one is only required to generate a reference profile and specify an axis of revolution, as shown in Fig. 5.9a. Once these elements are selected, the CAD software generates a revolution body, as shown in Fig. 5.9b.

It is important to remember that for parts that are 3D printed, it is required to consider a flat region that serves as the contact surface on which the 3D model can be printed. Given the above, as a rule, the reference profile must have a straight line at one of the profile ends. Allowing the generation of a base in the model (for the example presented, this line would be the one that generates the base of the pawn).

Additionally, employing the revolution of a profile, it is possible to generate bodies with some hole inside them (e.g., a toroid or doughnut), if the sketch to be revolted does not intersect the rotation axis of the operation.

An example of this type of bodies with central holes can be appreciated in Fig. 5.10, where a funnel is generated using only the side profile and the dimensions of the radius desired as a borehole.

On the other hand, revolution bodies can also perform subtractive operations such as the one shown in Fig. 5.11, where a rounding is applied to the corners of a nut to avoid sharp angles.

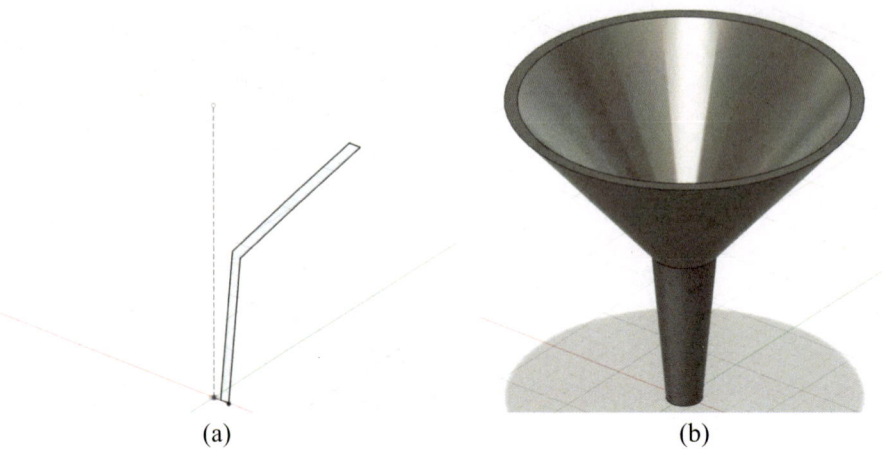

(a)   (b)

Fig. 5.10 Example of the use of the revolve tool for bodies with central holes. **a** Reference profile for the funnel. **b** Body generated after the application of the revolve operator

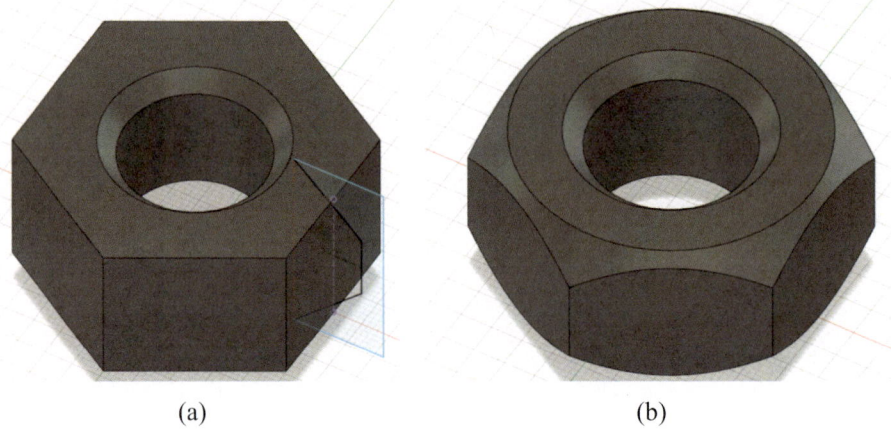

(a)   (b)

Fig. 5.11 Example of the use of the revolve tool as a subtractive operation. **a** Original model and profile for subtraction. **b** Final body with rounded edges

5.2.1.3 Sweep

The Sweep tool, although less frequently used, is also a helpful tool to consider, especially when designing structures that employ pipes or wires. Thus, this tool allows the generation of bodies that follow a guide path (usually employing a centerline but not limited to it) (Fig. 5.12).

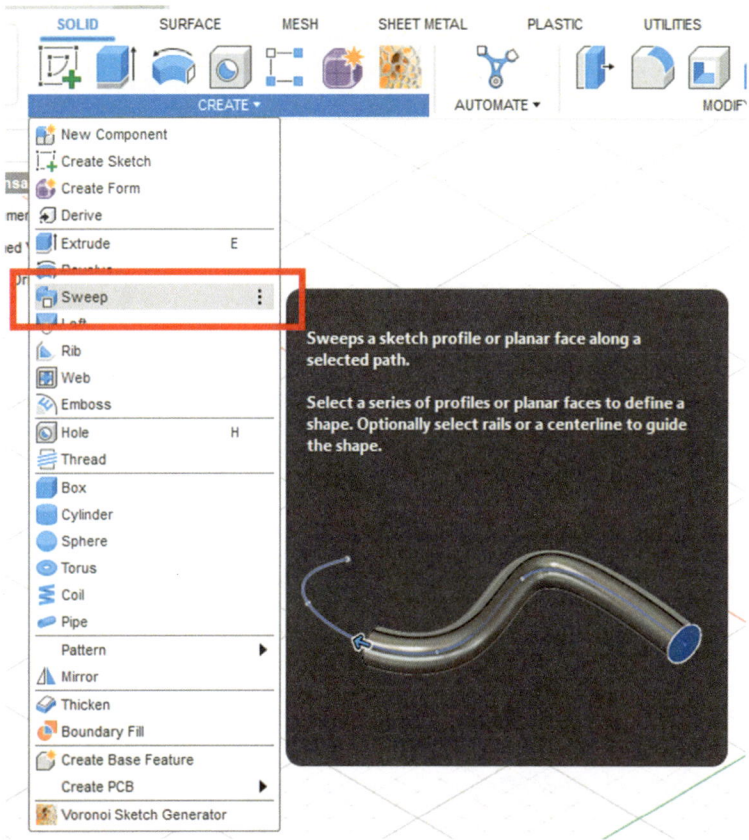

Fig. 5.12   Sweep tool in the drop-down menu of the CREATE panel

An example of the use of this tool is shown in Fig. 5.13, where it is possible to see how a 3D model of a wire toy can be generated (Fig. 5.13b) using only a profile with the wire cross-section and a guide path with the shape to be molded in the wire (Fig. 5.13a).
.

It is important to point out that the CAD software tries to solve as much as possible the discontinuities generated at the intersections of the edges of the base path, generating sharp interconnections between cylinders (as in Fig. 5.13b). However, if a design with contoured edges is required, it would be required to generate contoured curves in the guide path, eliminating any sharp edges.

Finally, the pop-up window of this tool also offers the flexibility to include some taper angles (similar to positive or negative extrusion angles), or twists (rotating the profile around the guide path). An example of the use of such parameters is presented in Fig. 5.14, where circular profiles (Fig. 5.14a) are extruded vertically while, at the same time, they are twisted and slimmed around the guide path (Fig. 5.14b).

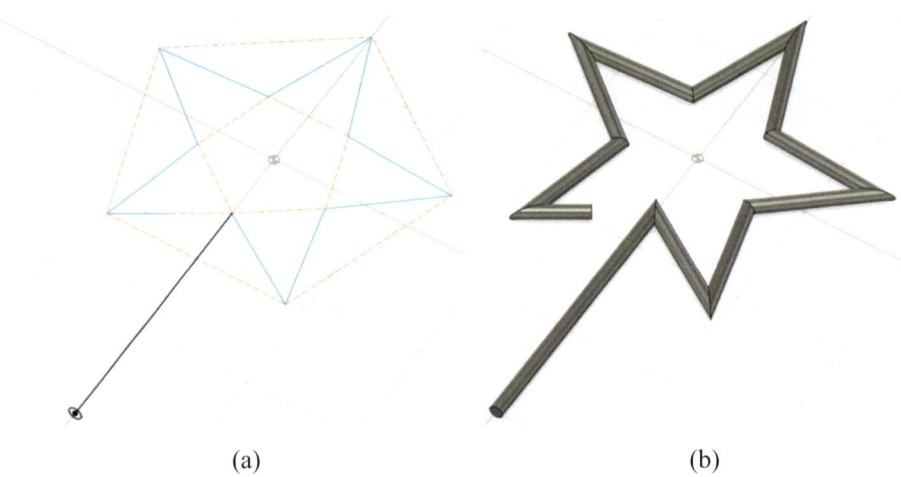

(a)   (b)

Fig. 5.13 Example of the use of the sweep tool to follow a path with a profile. **a** Guide path and profile used. **b** Body obtained after the use of the tool

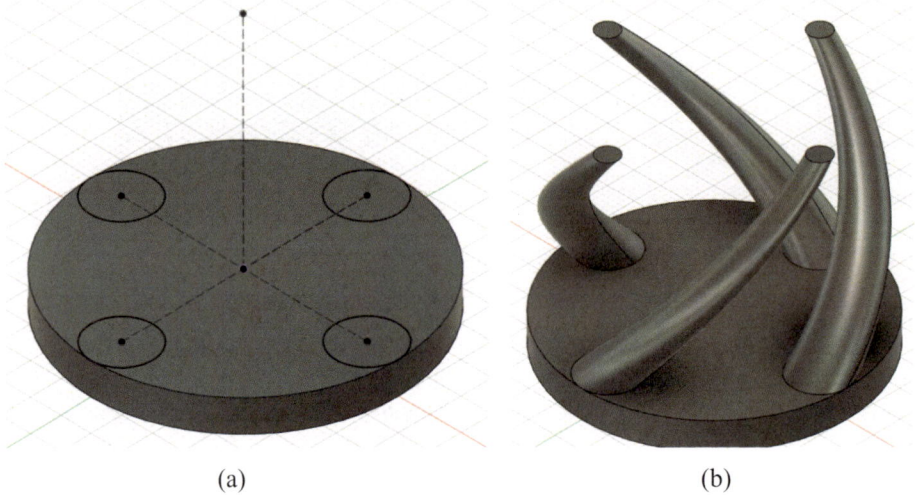

(a)   (b)

Fig. 5.14 Example of the use of the twist and taper angle parameters of the sweep tool. **a** Reference model with the required profiles and guide path. **b** Body obtained after the use of the tool

5.2.1.4 Thread

The Thread tool (Fig. 5.15) is a tool that allows the addition of threads to a hole on a model, using an extensive library of parameters that enable the generation of different types of metrics or standards.

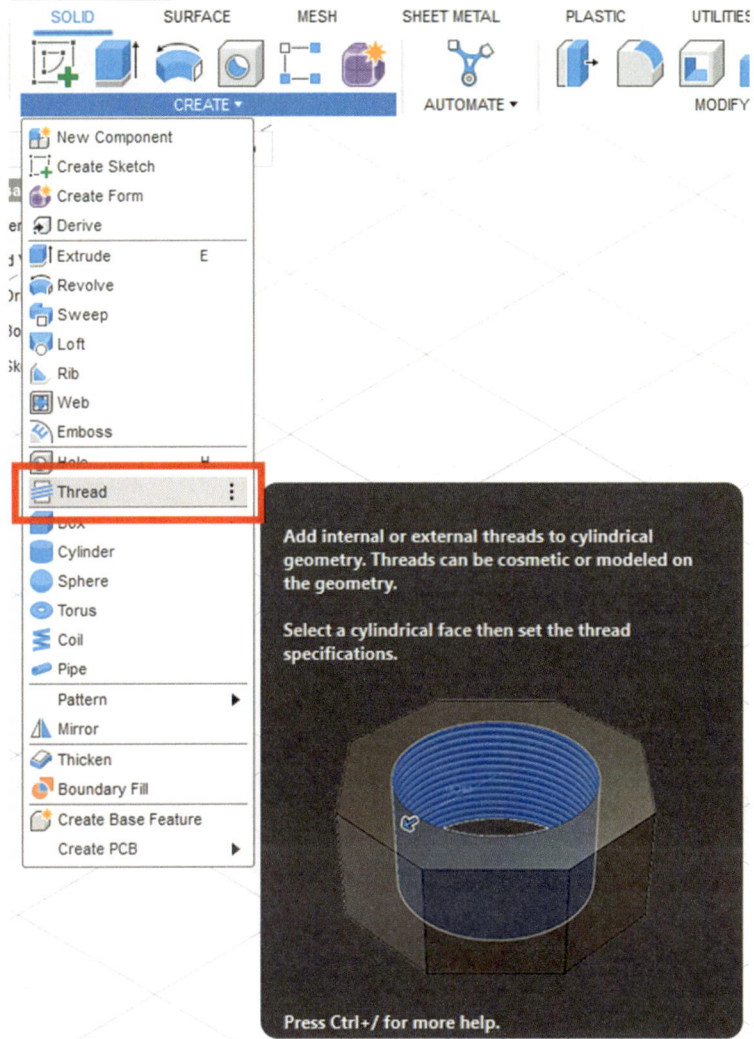

Fig. 5.15   Tread tool in the drop-down menu of the create section

The use of this tool greatly simplifies the design of a thread. While only being required a couple of clicks and the selection of a cylindrical surface and the desired standard to thread a model (Fig. 5.16), regardless of the original size of the cylindrical hole. However, although this tool has many features, it is also important to consider that its application to 3D printing has limitations due to the resolution that 3D printing can offer.

Fig. 5.16 Example of the use of the Tread tool. **a** Original 3D model. **b** Parameters and settings used. **c** Resulting model

Given the above, it is appropriate to evaluate whether the threading of interest is feasible to obtain by 3D printing or if it is more appropriate to use metal inserts with threadings and then thermally fuse them to the 3D print.

Finally, checking the "Modeled" box is vital when configuring the threading characteristics. Otherwise, the software only produces an apparent thread that does not represent a real geometry and will not be printed on the 3D printer.

5.2.1.5   Circular and Rectangular Patterns

Finally, it is usual that in addition to generating new features for the model, a pattern of already designed features is made in the model. Fusion 360 offers two options for it: circular and rectangular pattern tools (Fig. 5.17). Such tools allow the generation of a pattern around an axis or consider displacements on the axes of a plane.

Using these patterns, it is possible to select the type pattern and the target feature (either by manual selection of its faces or by selecting steps on the time sequence of operations at the bottom of the screen) and then adjust the axes of the operation and the number of its repetitions (Fig. 5.18).

Additionally, due to computational properties, it is usually considered that 3D models with 3D patterns have a better computational performance than equivalent 3D models generated with 2D patterns. Thus, using 3D patterns is recommended over their 2D sketch counterpart if possible.

Finally, it is a good practice to look at the rest of the operations in the drop-down menu of the CREATE section and be aware of the available 3D primitives or other tools in that section. Although such additional options may have infrequent use, they offer options that can significantly simplify several designs.

Fig. 5.17 Pattern menu in the
drop-down menu of the
CREATE section

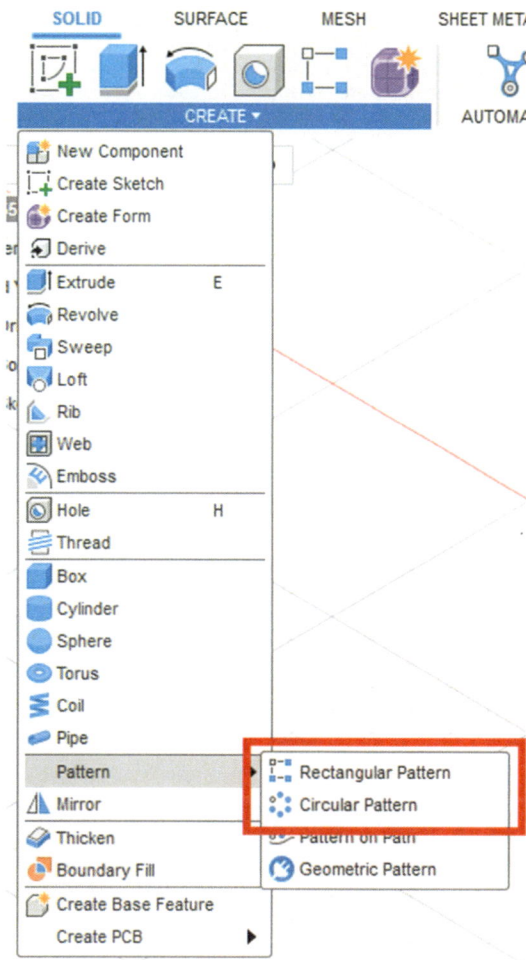

5.2.2 Panel Modify

As its name indicates, the MODIFY section (Fig. 5.19) contains most tools that allow
adding features to a 3D model, offering multiple tools that significantly reduce the time
required to add details to a 3D model.

5.2.2.1 Fillet

The Fillet tool (Fig. 5.20) is a flexible tool with several applications. It allows the smoothing of sharp angles in the model to generate rounded edges in their place. Thus, this
operation removes material at convex regions and adds material at convex regions.

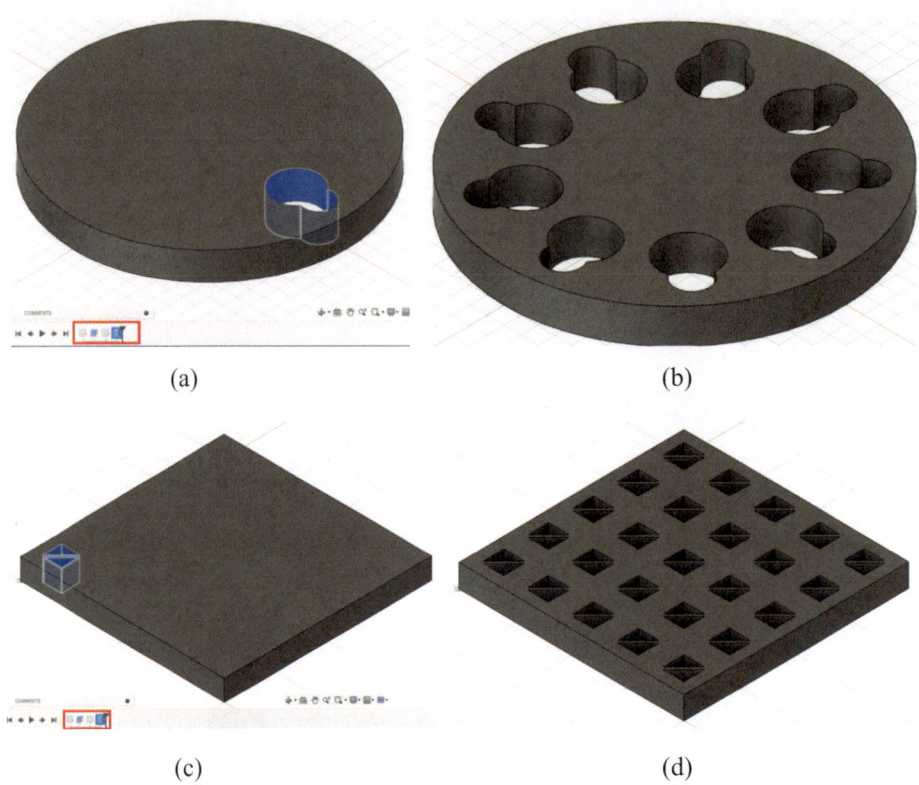

(a)   (b)

(c)   (d)

Fig. 5.18 Example of the use of circular and rectangular patterns. **a** Reference circular feature. **b** Model with the circular pattern applied. **c** Reference rectangular feature. **d** Model with the rectangular pattern applied

Fig. 5.19 MODIFY panel on the contextual strip SOLID

Fig. 5.20 Fillet tool in the MODIFY section

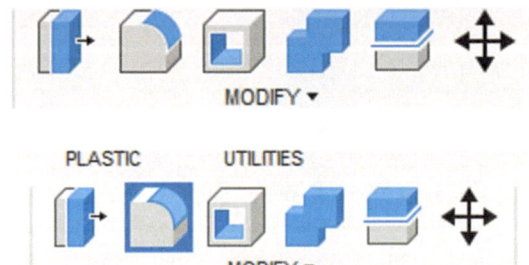

An example of the use of this tool can be seen in Fig. 5.21, where it is used to generate rounded edges that are comfortable to hold in the handle of a hand tool. In this example, it can be seen how a base shape composed only of straight surfaces is used. However, this modifier makes obtaining a more comfortable and appropriate shape possible by adding or removing material in the required regions.

Concerning 3D printing, this tool can also be used to widen the connection points between various parts of a 3D model (similar to a rib/stiffener), thus generating elements that are more resistant to lateral forces at the connection points, as shown in Fig. 5.22.

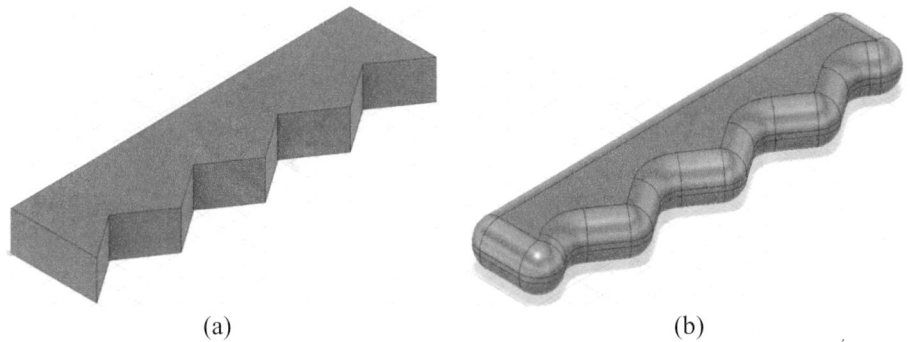

(a)   (b)

Fig. 5.21 Example of the use of the Fillet tool. **a** Original 3D model. **b** 3D model resulting from the application of the Fillet tool

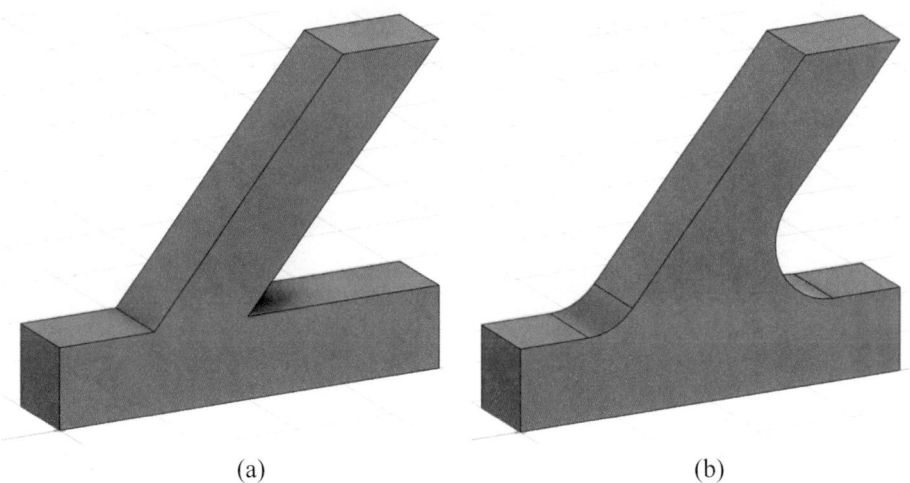

(a)   (b)

Fig. 5.22 Example of using the Fillet tool to generate reinforcements in model interconnections. **a** Original 3D model. **b** Reinforced 3D model

5.2.2.2 Chamfer

The chamfer tool (Fig. 5.23) offers similar behavior to the one generated by the Fillet tool, with the difference of generating straight cuts instead of rounded corners. This property removes steep angles from geometries and substitutes them with two swallow angles.

On the other hand, analogously to Fillet reinforcements, this tool can also generate ribs in the connections between segments, increasing their strength, as shown in Fig. 5.24.

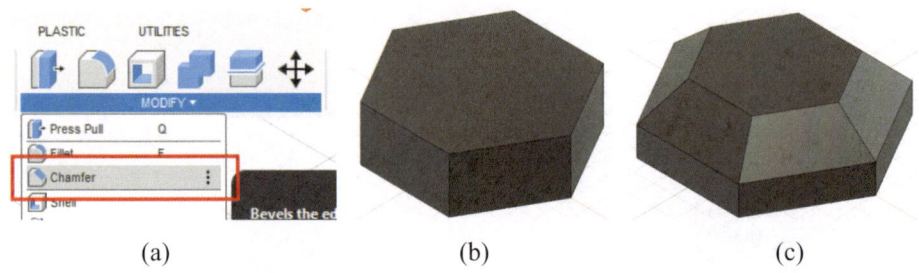

| (a) | (b) | (c) |

Fig. 5.23 Location and example of use of the Chamfer tool. **a** Location of the tool on the MODIFY drop-down menu. **b** Original 3D model. **c** Modified 3D model

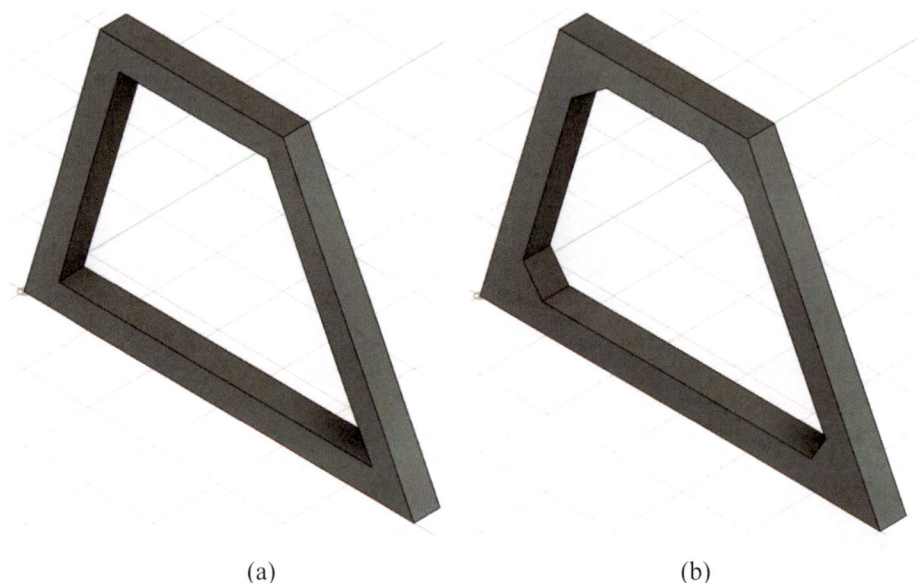

| (a) | (b) |

Fig. 5.24 Example of the use of the Chamfer tool to generate edge reinforcements. **a** Original 3D model. **b** Reinforced 3D model

Nevertheless, it is important to point out that other Fusion 360 tools also specialize in the design of reinforcements, such as the Rib tool in the CREATE section, which focuses explicitly on this task.

5.2.2.3   Shell

A relatively helpful tool is the shell tool (Fig. 5.25), which allows the hollowing of a model while preserving a constant thickness in the walls of the model. This automatic hollowing makes it easier to generate shells or parts with some specified thickness, even if they have complex shapes.

An example of the above can be seen in Fig. 5.26, where a vessel with a hexagonal shape that simultaneously has a twist is generated. To generate the wall of such a vessel with a constant thickness, it is only necessary to select the upper face of the vessel (Fig. 5.26a) and the shell modifier, to finally indicate the desired thickness. After that, a resulting vessel as the one in Fig. 5.26b would be obtained. Concerning the wall thickness, a section analysis of the vessel is presented in Fig. 5.26c, showing that it has a constant thickness regardless of the complexity of the geometry of the vessel.

Fig. 5.25 Shell tool in the MODIFY section

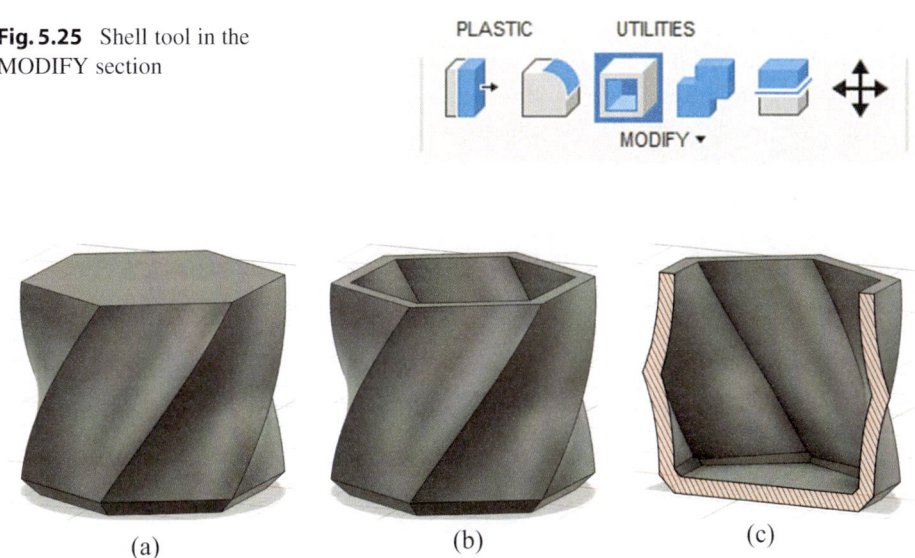

(a)   (b)   (c)

Fig. 5.26 Example of the use of the Shell tool to generate a model with walls of a specified thickness. **a** Original 3D model. **b** Hollowed out 3D model. **c** Section analysis of the generated vessel

5.2.3   Panel Assemble

The assemble section (Fig. 5.27) provides a set of tools for visualizing and modeling the behavior of an object composed of several components. The most essential tools of this section are the Join and Motion Link tools.

5.2.3.1   Joint

Within 3D design, multiple parts are expected to be used to construct a machine or object. For such scenarios, it is only necessary to generate a new file, save it, and then add all the parts needed for the model by dragging them to the work canvas. Once the parts of interest are found in the canvas, it is possible to use the Joint tool of the ASSEMBLE section (Fig. 5.28). This tool allows the establishment of movement constraints between the parts of the model, so they can generate the mechanical behaviors that would be expected in the real world.

To simulate complex behaviors, the Joint tool includes among its options the ability to generate 7 different types of connections, which are shown in detail with an illustrative example in Table 5.1.

Thus, using such constraints, it is possible to generate complex mechanical links, such as the pantograph shown in Fig. 5.28, which allows a draw to be enlarged (Fig. 5.29b, red arrows) using a reference marker (Fig. 5.29a, black arrows).

On the other hand, in the case of angle-to-angle or angle-to-displacement motion relationships, it is possible to use a joint as a motion base and then establish the specific relationship by means of a "Motion Link."

5.2.3.2   Motion Link

The Motion Link tool (Fig. 5.30), is a tool that allows establishing motion relationships between components that interact through some kind of contact/friction (such as gears or gear racks) and not through piece anchors (such as the previously presented pantograph).

To make use of this tool, it is only necessary to select the joints to be linked and, depending on the type of displacement, indicate the parameters that direct the movement.

Fig. 5.27 ASSEMBLE panel on the contextual strip SOLID

Fig. 5.28 Joint tool in the ASSEMBLE section

Table. 5.1   Mechanical constraints of the Joint tool with icons and examples

Restriction name	Icon	Characteristics	Example
Rigid		This constraint is used when desired to generate a solid connection between materials, subordinating the movement of one part to the other	The welding of two metallic objects
Revolute		This restriction allows a part to rotate around but not along an axis	A wheel on a toy car
Slider		This constraint allows an object to slide on one axis but prevents it from moving on the other	The movement of a drawer
Cylindrical		This constraint is similar to the Revolute constraint, but unlike the Revolute constraint, the Cylindrical constraint does allow movement along the axis	The head of a soap dispenser
Pin-Slot		This constraint allows a body to slide in a slot while allowing the body to rotate	The movement of a securing screw on a groove
Planar		This type of constraint allows a body to slide on the two axes of a surface	The movement of a book sliding on a table
Ball		This type of constraint allows a body to rotate about a spherical fulcrum	A ball join in a car direction system

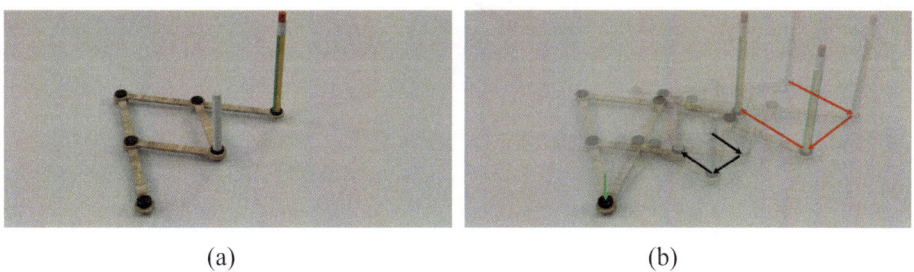

(a)   (b)

Fig. 5.29   Example of the linking of several bars to generate a pantograph. **a** Rendered image of the pantograph with its ten components. **b** Superimposed images showing the mechanical linkage of the pantograph used to replicate shapes at different scales

For example, Fig. 5.31 shows two gears with different numbers of teeth (24 and 12, respectively). Therefore, a direct 1-to-1 rotation would not reflect an appropriate behavior for this system. Given the above, it is possible to correctly link both revolutions by indicating that both gears must rotate around their axis using the revolute option of the

Fig. 5.30 Motion link tool in the drop-down menu of the ASSEMBLE section

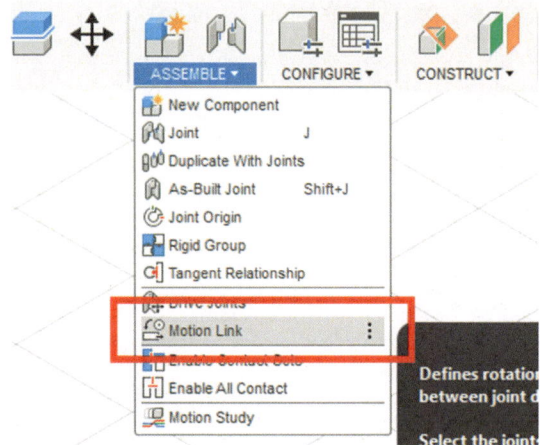

joint tool and then create a motion link, where for every 360° of rotation of the large gear there will be 720° of rotation in the small gear.

Once this relationship is established, it is possible to rotate one of the gears, and its counterpart will rotate accordingly, either reducing or increasing the rotation speed.

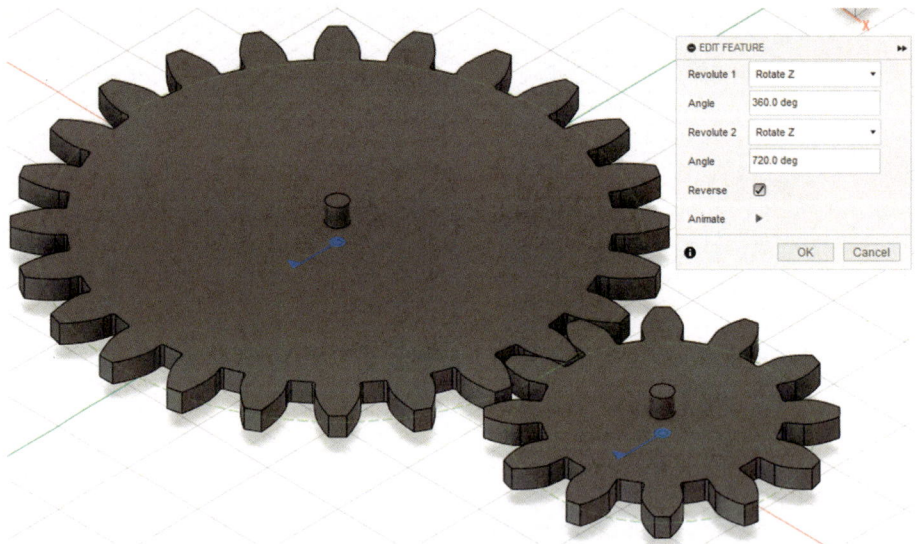

Fig. 5.31 Example of the use of the Motion Link tool for linking two gears with different amounts of teeth

5.3   3D Design Exercise

As shown throughout the chapter, successful 3D design involves the ingenious use of tools to express the characteristics of a 3D model most appropriately and conveniently. To do this, it is necessary to analyze the body to be modeled and deconstruct it to its most basic foundations. With such a procedure, it is possible to quickly identify the elements to be modeled and visualize the most appropriate design procedure.

In order to encourage the development of these skills, it is convenient to model everyday objects that one may have at home, seeking to apply the broadest variety of operations in the process. To demonstrate this skill development exercise, the modeling process of the coffee table presented in Fig. 5.32 is presented in the following sections.

5.3.1   Table Support Design

Initially, it is possible to start the table design with its support structure (Fig. 5.33), which is fabricated by meticulously bending a circular profile. Analyzing this bending process, it is possible to suggest that a line route and the application of a sweep operation would offer a straightforward design process.

On the other hand, although modeling this part requires using a 3D sketch, it is designed in three parts to maintain simplicity and demonstrate the use of assemblies.

Initially, the base of the support can be designed, generating a guide route as shown in Fig. 5.34a, which is a rounded version with 10 cm Fillets of the straight lines of Fig. 5.34b.

Subsequently, in an additional sketch, it is possible to specify the profile to be used in the sweep operation, in this case, a 10 mm circular profile (Fig. 5.35a). Once the

Fig. 5.32 Modern coffee table. Result of the modeling and assembly of the skill development exercise

Fig. 5.33 Modern coffee table
stand

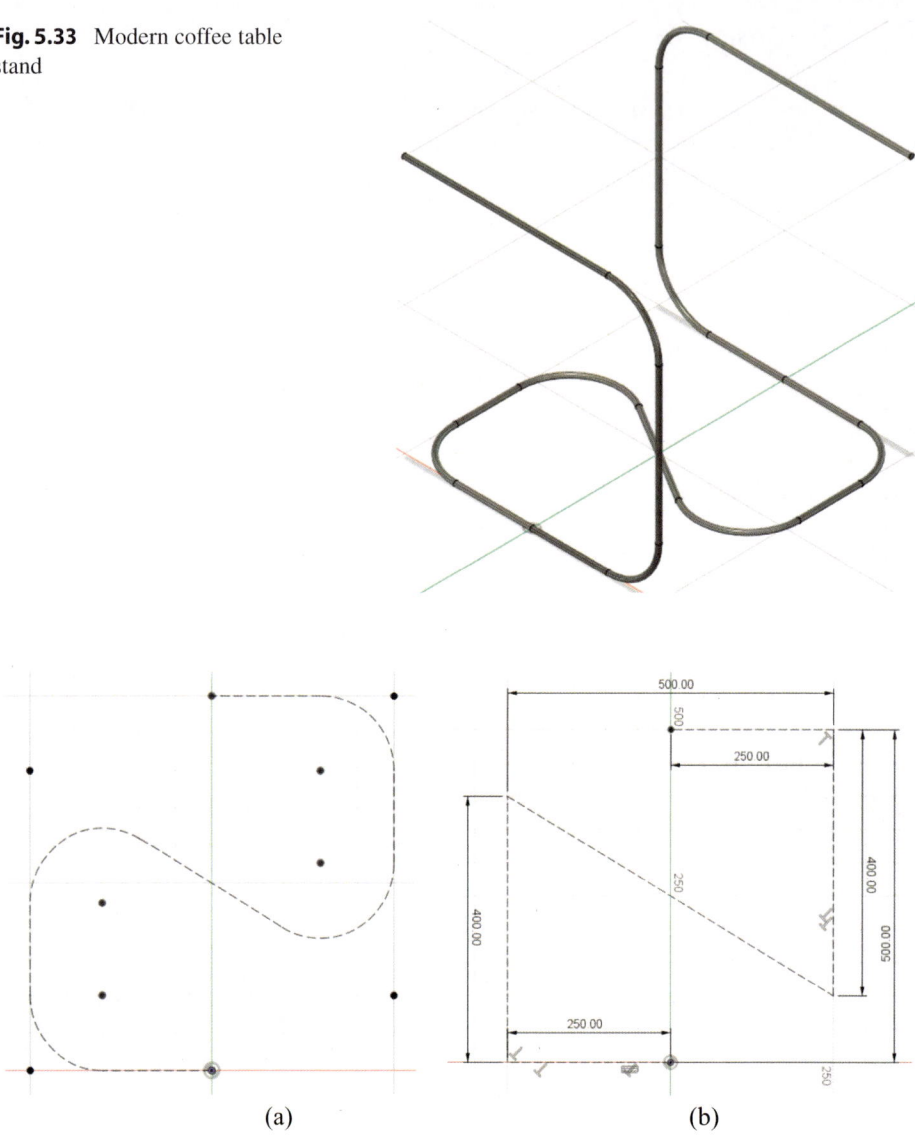

Fig. 5.34 Guide path of the base of the table support. **a** Final guide path. **b** Guide path before
rounding the edges

operation has been applied, the first piece of the table support should be obtained, as
shown in Fig. 5.35b.

Subsequently, to finish the design of the pieces of the table support, the circular profiles
of Fig. 5.36a must be designed, which are built in a similar way to the first base but with
the dimensions of Fig. 5.36b and with roundings of 10 cm radius in its edges.

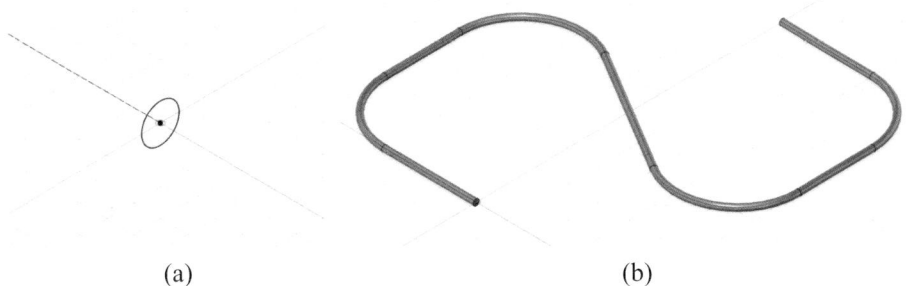

(a)	(b)

Fig. 5.35 First part of the coffee table support. **a** Path and profile used in the sweep operation. **b** Resulting support

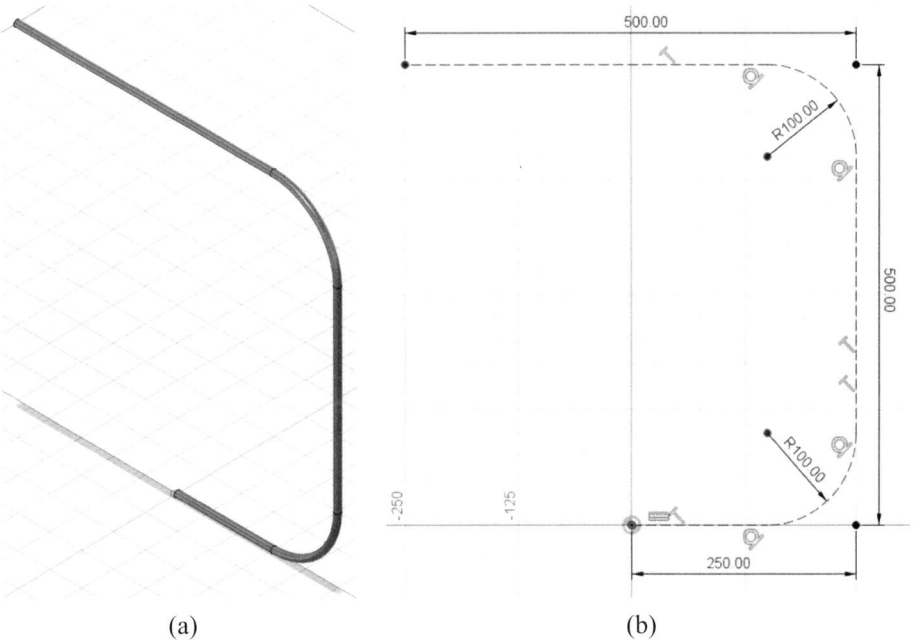

(a)	(b)

Fig. 5.36 Lateral supports of the coffee table. **a** 3D model of the support. **b** Guide path used in the support

Finally, it is possible to complete the construction of the table support by making an assembly where the parts are positioned and joined with rigid joint constraints. This operation can be seen in Fig. 5.37a, while the complete arrangement can be seen in Fig. 5.37b.

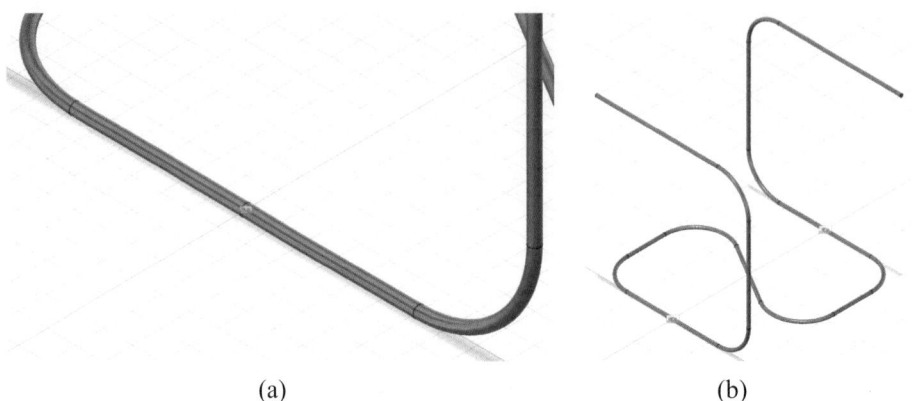

<table>
<tr><td>(a)</td><td>(b)</td></tr>
</table>

Fig. 5.37 Coffee table support. **a** Close-up to the side support junction. **b** general view of the 3D model of the coffee table support

5.3.2 Tabletop Design

Once the table support is available, it is possible to design a wood block that serves as the main surface of the table. For it, it is possible to design a 75 cm × 75 cm square in a sketch in the XY plane (Fig. 5.38a), after which it is possible to extrude the square 1 inch (Fig. 5.38b).

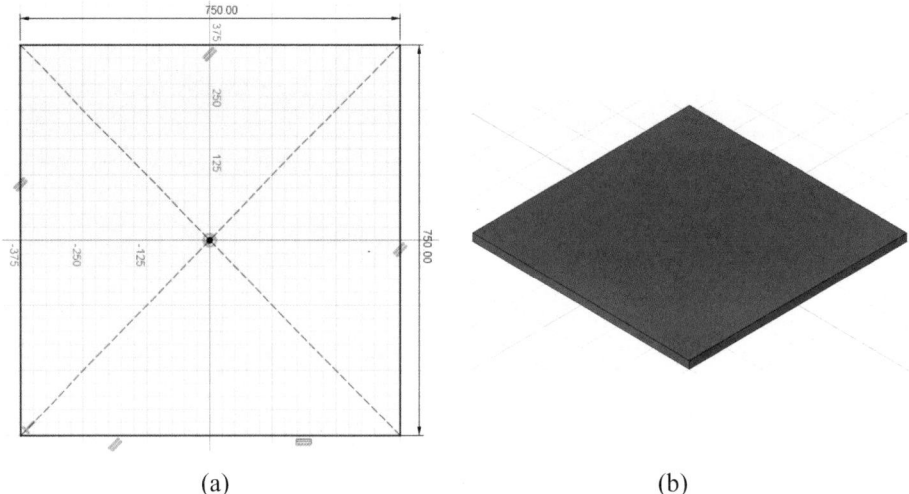

(a)   (b)

Fig. 5.38 Main body of the surface of the table. **a** 2D sketch with the dimensions of the table. **b** Base body of the main surface of the table

Subsequently, to offer a more contoured design, the corners of the table are rounded, considering a 15 cm fillet (Fig. 5.39a). A 20 mm chamfer is applied (Fig. 5.39b) to all the sides of the table. Finally, a rounding of 4 mm is applied to the table's sharp corners (Fig. 5.39c).

On the other hand, it is also necessary to generate a set of threaded holes and their reflections in the lower part of the table (note that screw threadings are used as a didactic example for the use of threads since wood would traditionally use wood screws), to insert M6 bolts that hold the tabletop to the table supports. This can be achieved by using the extrude tool on a set of circles considering a depth of 20 mm (Fig. 5.40a) and then applying an M6 thread using the Tread tool.

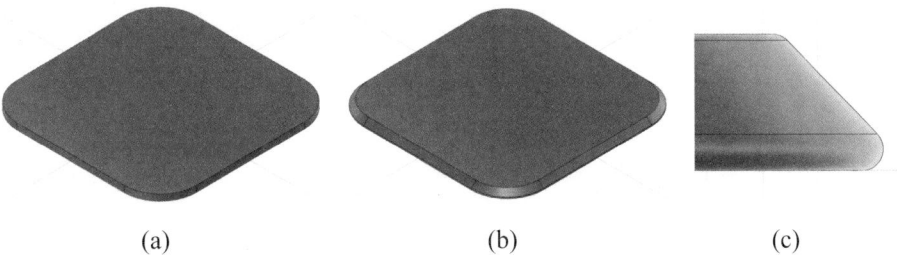

(a)   (b)   (c)

Fig. 5.39 Rounding of the table. **a** Corner rounding. **b** Chamfering of the edges. **c** rounding of the whole body

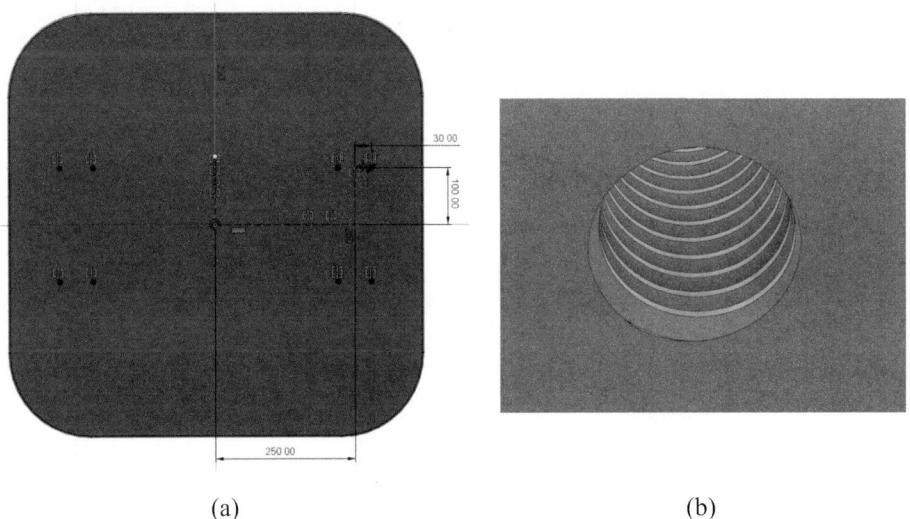

(a)   (b)

Fig. 5.40 Set of threaded holes to secure the tabletop to its support. **a** 2D sketch with the position of the threaded holes. **b** Threaded hole with an M6 profile

Fig. 5.41 Main surface of the
table

Finally, using the hotkey "A," a texture can be selected to better emulate the material
used for this part, allowing the table surface shown in Fig. 5.41 to be obtained.

5.3.3   Design of the Table Anchors and Screws

Finally, to fix the table to the supports, it is possible to use a set of anchors and screws
such as those shown in Fig. 5.42.

To generate the proposed anchors, it is only necessary to extrude a side profile
(Fig. 5.43a) that allows for tightening the supports' profiles to the table's main surface.
Such a structure can be efficiently designed by considering a set of reference construction
lines that subsequently allow the application of an offset modifier with the thickness of a
standard metal sheet.

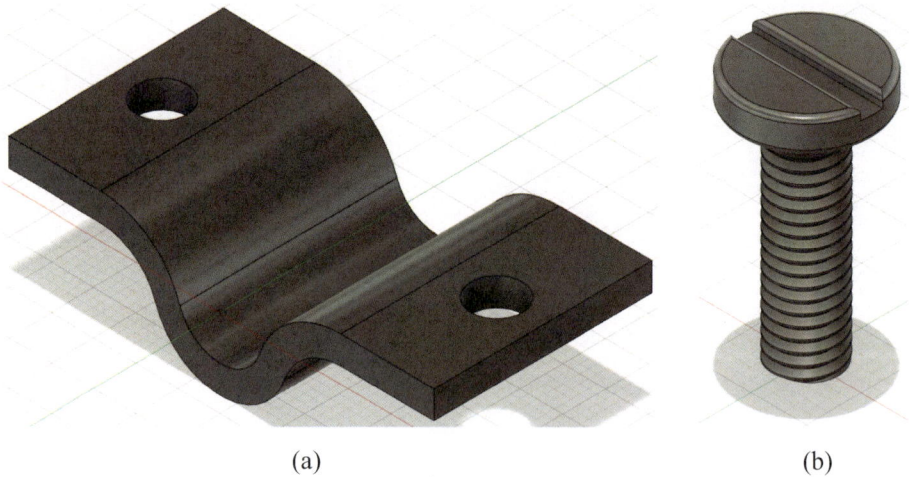

             (a)   (b)

Fig. 5.42 Anchors and screws used to fasten the tabletop to its supports. **a** Metal anchors **b** Screws

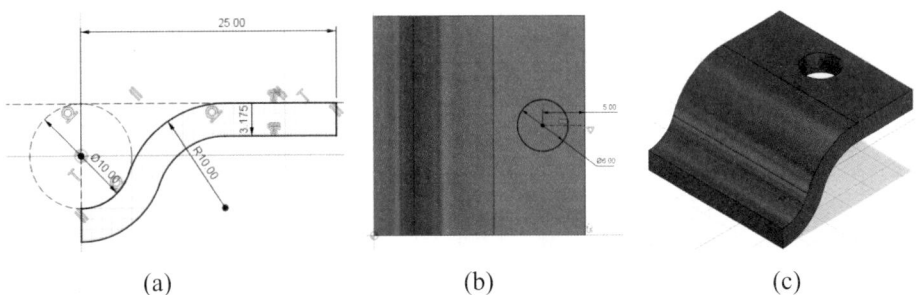

(a)   (b)   (c)

Fig. 5.43 Design process of the metal anchorages. **a** Half lateral profile. **b** Generation of guide sketches for holes. **c** Half metal anchor

Subsequently, it is necessary to generate a set of holes that allow the screws to be positioned appropriately. To do this, it is possible to generate a circle in the flat region of the support and then apply a subtractive extrusion that generates a hole, as shown in Fig. 5.43b and c. Finally, it is possible to apply a mirror modifier on the half anchor to obtain its counterpart and complete the 3D model (Fig. 5.42a).

Regarding the screws, although there are multiple ways to generate them, in this example, for didactic reasons, they are designed using a body of revolution as shown in Fig. 5.44a, which allows generating a base body (Fig. 5.44b) quickly and effectively. Once such a base body is available, it is possible to add an M6 thread (Fig. 5.44c) and a notch (Fig. 5.44d) on the screw head. Finally, it is possible to add a set of roundings (Fig. 5.44e) to complete the screw construction.

5.3.4   3D Model Assembly

Once all the parts used in the 3D model are available, assembling them with joints is possible. However, given the repetition of several modules (metal anchors and screws), it is more convenient to generate an auxiliary assembly that already contains all these assembled parts, thus avoiding having to indicate the references of each element.

Given the above, Fig. 5.45 shows such an arrangement, which has been achieved by referencing the lower faces of the screw heads to the holes of the flat surface of the metal anchors employing a rigid joint.

For aesthetic purposes, it may be advisable in some scenarios to hide the joint symbols inside the 3D model (Fig. 5.46). To achieve this goal, one may click on the eye next to the Joints folder to hide them (Fig. 5.46a and c). If required, they can be made visible again by pressing that same location (Fig. 5.46b and d).

Once the anchor and screw arrangements are in place, it is possible to generate a new file where the table surface is added and the four necessary anchors are added to its

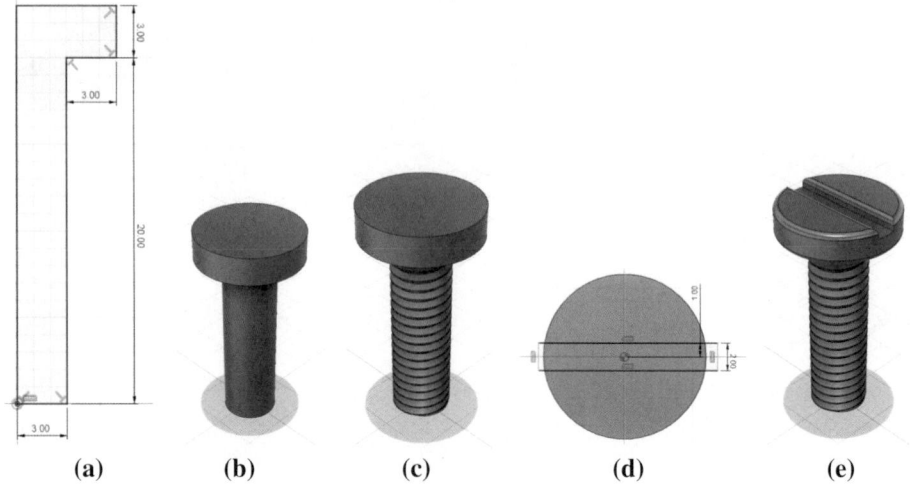

Fig. 5.44 Screw design process. **a** Screw revolution profile. **b** Base body after revolution. **c** Screw threading. **d** Sketch for generating the Screw notch. **e** Rounding of the screw contact edges

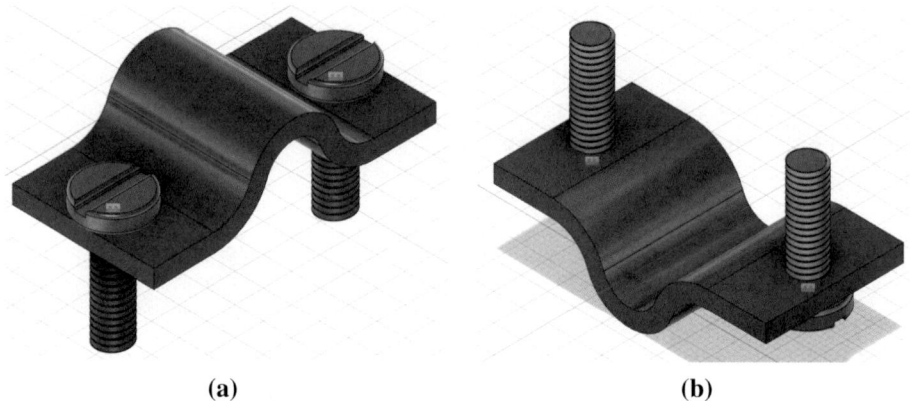

Fig. 5.45 Assembly of metal anchors and screws. **a** Upper isometric view. **b** Lower isometric view

bottom surface, joining them one by one in their respective positions through rigid joints, as shown in Fig. 5.47.

On the other hand, once the anchors are in their proper positions, the table support can be added to the file (Fig. 5.48a) to finish the assembly. Since there is no direct position reference, one of the two edges of the support profile must be selected and constrained to be 137.3 mm offset from one of the edge anchors, as shown in Fig. 5.48b.

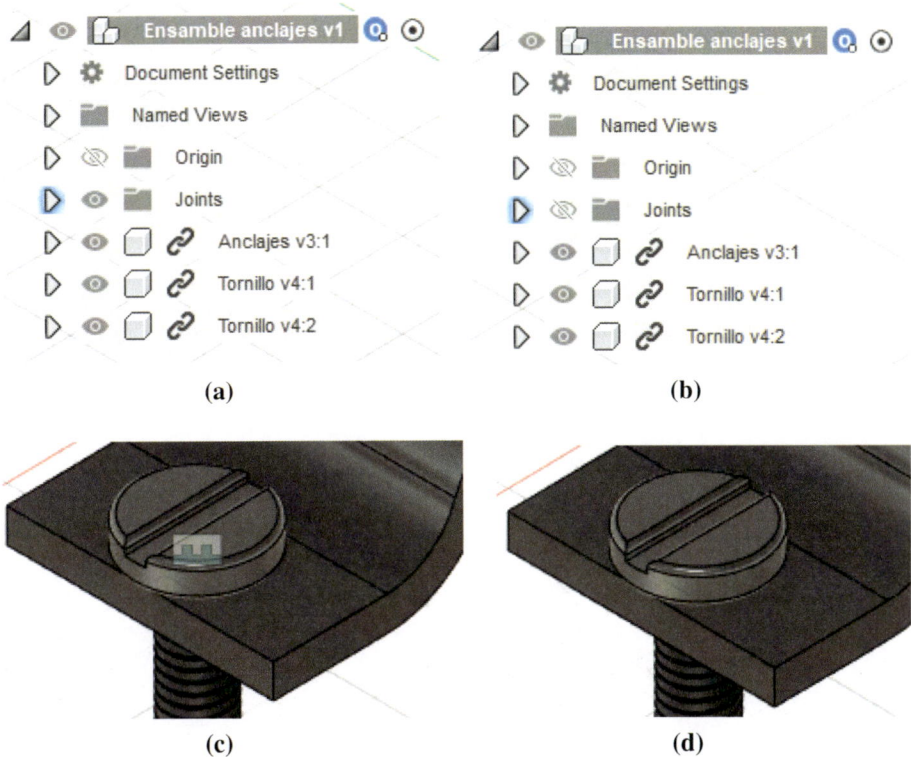

Fig. 5.46 Use of the side marker in the Joints folder to switch between visible and non-visible states of the joints. **a** and **c** Visible state in the joints. **b** and **d** Non-visible state in the joints

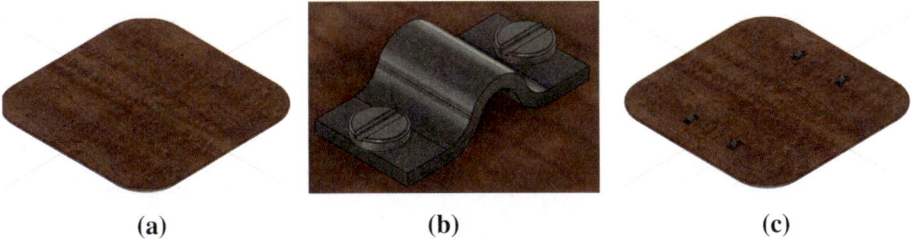

Fig. 5.47 Assembly of anchors and screws to the table bottom surface. **a** Table bottom surface. **b** Example of complete anchor assembled. **c** Table bottom surface with metal anchors installed

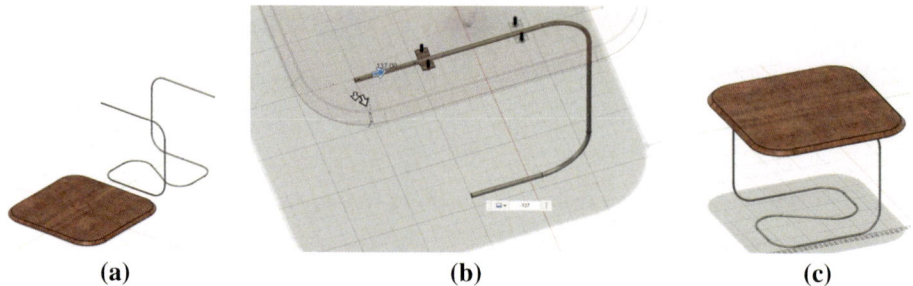

(a)   (b)   (c)

Fig. 5.48 Assembly of the main surface to the table support. **a** Elements to be assembled. **b** Anchoring of the support to the table. **c** Assembly result

(a)   (b)

Fig. 5.49 Renders of the example table. **a** Top perspective view of the model. **b** Bottom perspective view of the model

This constraining allows the support to be positioned in the center of the table, keeping the table symmetric.

Finally, once the assembly has been finalized, it is possible to switch between the DESIGN environment and the RENDER environment in the left upper corner to generate a high-resolution image of the proposed example, as shown in Fig. 5.49.

Finally, as indicated at the beginning of this section, 3D design is a skill obtained through continuous and persistent practice, since it requires the designer to be aware of the essential nature and details that make up his model. Given the above, the readers are invited to model two or three other simple objects that they may have at home (household appliances, kitchen elements, furniture, lamps, etc.), to improve their design intuition and later be able to design more complex objects such as the design of 3D printed electric motors.

References

1. O. Kontovourkis and G. Tryfonos, Integrating parametric design with robotic additive manufacturing for 3D clay printing: An experimental study. In ISARC 2018—35th International Symposium on Automation and Robotics in Construction and International AEC/FM Hackathon: The Future of Building Things, International Association for Automation and Robotics in Construction (I.A.A.R.C) (2018). https://doi.org/10.22260/isarc2018/0128

2. B. Rădulescu et al., Thermal expansion of plastics used for 3D printing. Polymers (Basel) **14**(15) (2022). https://doi.org/10.3390/polym14153061

3. N. Zhang, L.-C. Zhang, Y. Chen, Y.-S. Shi, Local barycenter based efficient tree-support generation for 3D printing ☆,☆☆. Comput. Aided Des. **115**, 277–292 (2019). https://doi.org/10.1016/j.cad

4. H.D. Budinoff, S. McMains, Will it print: a manufacturability toolbox for 3D printing. Int. J. Interact. Des. Manuf. **15**(4), 613–630 (2021). https://doi.org/10.1007/s12008-021-00786-w

5. B. Rankouhi, S. Javadpour, F. Delfanian, T. Letcher, Failure analysis and mechanical characterization of 3D printed ABS with respect to layer thickness and orientation. J. Fail. Anal. Prev. **16**(3), 467–481 (2016). https://doi.org/10.1007/s11668-016-0113-2

Setting up a 3D Model for a 3D Printer

6

6.1   General Overview of the Complete 3D Printing Process

In previous chapters, the base foundations for generating 3D designs that effectively reflect the desired features of an object were presented. However, the file associated with such a 3D model is in a format that 3D printers, by nature, cannot understand. Thus, a preprocessing of such a file is required to generate a set of instructions that the 3D printer can follow.

An illustrative diagram of this process is shown in Fig. 6.1. In the first step of such a diagram, it can be seen the process that must be followed to convert the 3D model in the CAD design software into a .stl (Standard Tessellation Language) file for its printing. This process is required to allow the 3D model to be described by a set of triangles that only describe the shape of the body and eliminate any color characteristics, visual texture, or material property.

Subsequently, this file must be processed by a layer processor, usually known as a "slicer," which combines the information of the 3D model (shape) with a set of printing parameters (desired properties) that alter the specific way of printing of the model. This stage is the most discussed in this chapter since it has determining implications for the final characteristics of the printed object. During the slicing stage, the density, material distribution, or even the type of filler to be used in the internal volume of the 3D object can be altered, leading to better or worse mechanical properties in the object.

Finally, the file must be sent to a 3D printer according to the characteristics of the material and sizes considered in the slicer. Regarding this stage, some tips & tricks are presented to simplify the printing process.

© The Author(s), under exclusive license to Springer Nature Switzerland AG 2025   157
E. Cuevas et al., *DC Motors*, Synthesis Lectures on Engineering, Science, and
Technology, https://doi.org/10.1007/978-3-031-64354-5_6

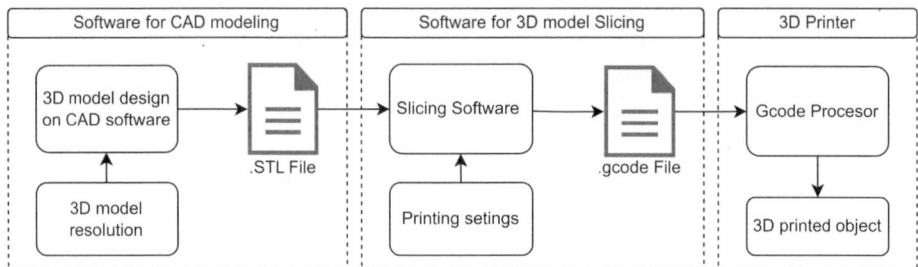

Fig. 6.1   Diagram of the complete 3D model printing process

6.2   Generating an .STL File

The first step in 3D printing is exporting the 3D model in a standard format for a 3D body description. Currently, the .STL format is considered the broadest standard format for this task. This file converts the original 3D body into a set of triangles that approximate the initial model together. However, as can be easily inferred, the transformation from a continuous body to a polygonal body has some distortions.

In order to mitigate distortions in the 3D model, the export menus usually request a target resolution level to be used, thus allowing a balance between the computational weight generated by having many vertices (and, therefore, more resolution) and the loss of quality inherent in the conversion.

In the case of Fusion 360, the export of a .STL file can be achieved by selecting the body to be exported and then right-clicking it to display a menu where the option "Save As Mesh" (Fig. 6.2a) can be selected. This action deploys an auxiliary menu (Fig. 6.2b) in the right section of the screen, where one can select the desired format, the base unit of the file, and the quality level (in this case, High). Although there are some additional advanced options, the default values provided by Fusion 360 tend to offer good results in most scenarios.

6.3   Generating a .Gcode File

Once a tessellation file is available, the next task involves converting the 3D tesselated body into a set of instructions that a 3D printer can understand and obey. Such instructions include, for example, heating the extruder to some temperature, specifying the flow of material to be extruded, indicating the positions the printer should reach, as well as the speed at which the extruder head should move.

This set of instructions can be easily generated using some slicing software. In this book, the use of Ultimaker Cura is considered, as it is a free and easy-to-use 3D printing software that is widely accepted by the 3D printing community.

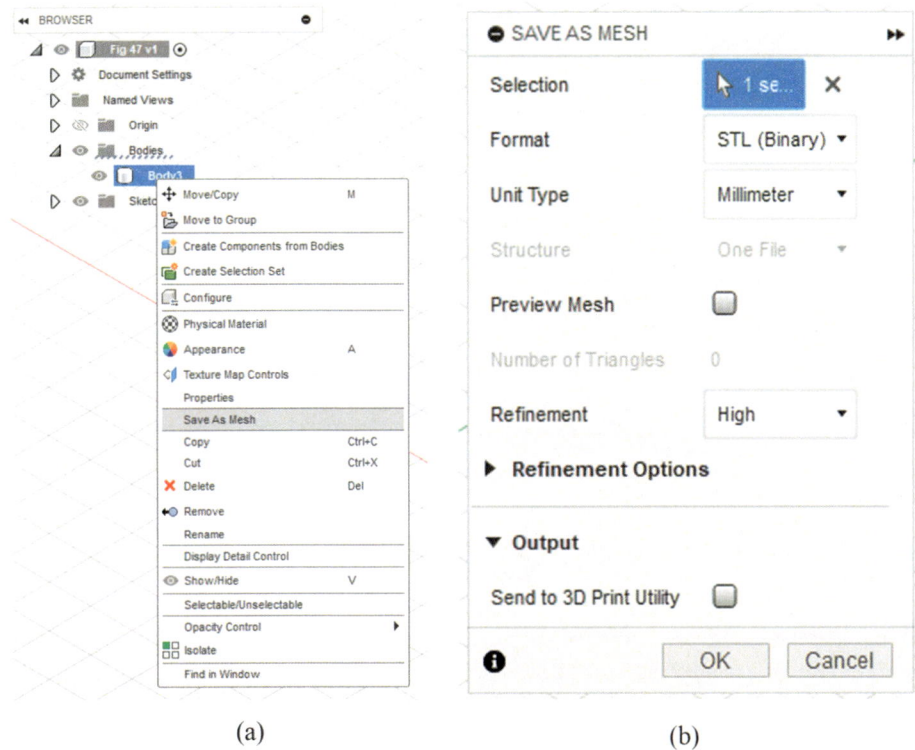

(a)   (b)

Fig. 6.2 Menus and submenus for exporting the 3D body as a mesh. (a) Selection of the body to be exported (b) Pop-up menu with export settings

To install Ultimaker Cura, one can access their website and download their installer. Then, after launching the installer, a couple of instructions are displayed through a user interface. Among such instructions, it is required to select the profile of the printer on which one desires to print (so that it preloads the manufacturer's default parameters).

Once the installation process has been completed, the program welcomes the user with a screen similar to the one shown in Fig. 6.3. In the present case, this screenshot shows a preview of the printing surface of the Flsun V400 printer. However, the printing area may have a different shape and dimensions if another printer is used.

This screen offers three main interface panels for configuring the printing process and a menu bar for program administration. The first print configuration section allows the management of different printers, the second one allows the management of different materials, and the last section allows the configuration of the printer parameters to fit the desired model.

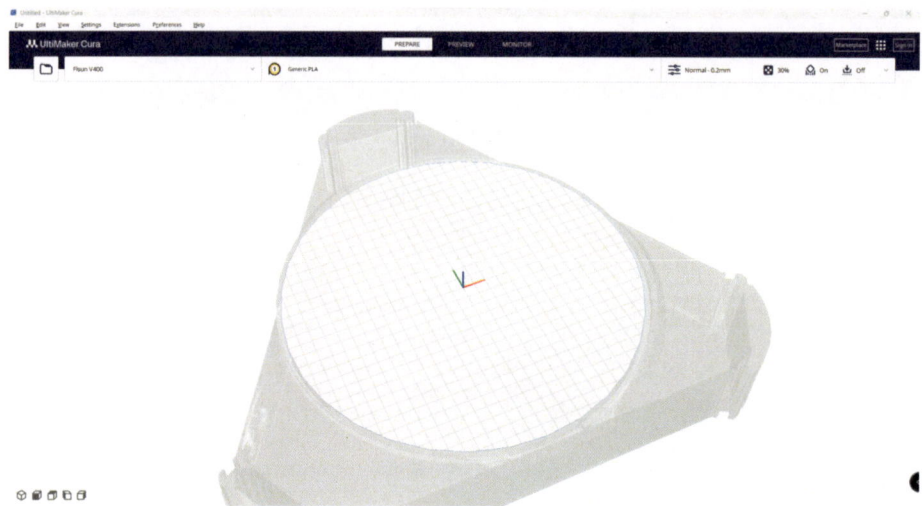

Fig. 6.3 Initial screen of the Ultimaker Cura software

6.3.1   Printer Management

The first interactive section (currently showing the name of the selected printer), allows a quick switch between different printer profiles (Fig. 6.4), thus facilitating the printing process in the case of having multiple printer models. For the following examples, the representative models shown in Fig. 6.4 have been selected because they allow the explanation of different properties of different kinds of printers.

Furthermore, in addition to switching between profiles, this section also allows the addition of new printer profiles, either by selecting one available in the printer library or

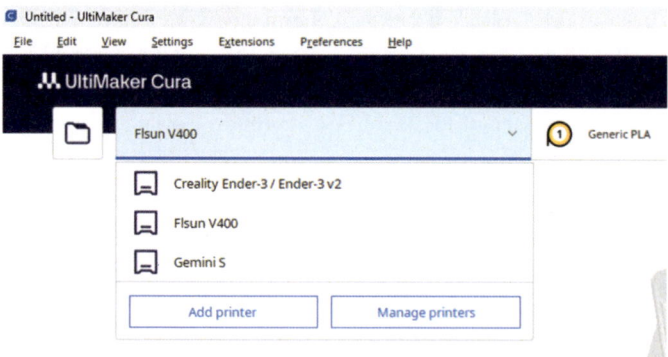

Fig. 6.4 Printer management panel

by manually setting each parameter (as in the case of the Gemini S printer). In the second case, it is only required to click "Add printer" and follow the instructions provided and the manufacturer's manual.

6.3.2   Materials Management

Similar to how printers are managed, it is also possible to use the second section of the print configuration panel (Fig. 6.5) to interleave between different materials or modify their properties. In the case of the example presented, a roll of PLA filament is selected for extruder 1. However, as shown, selecting from an extensive filaments library or generating a custom profile is also possible.

The appropriate selection of the printing parameters, considering the current state of the printing environment, is fundamental to achieving a good printing result. Given the above, locating the 3D printer in a room with a stable temperature is advisable, since

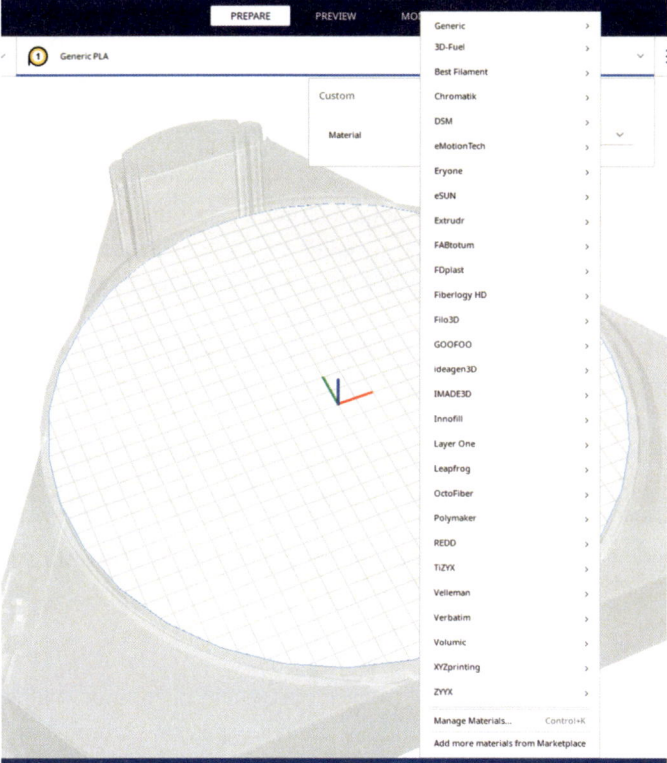

Fig. 6.5   Material management panel

such variations can noticeably affect the final print quality. For example, thermal variations during printing can generate a phenomenon known as warping, which consists of noticeable deformations in flat regions of the 3D printed model.

Moreover, variations in temperature or humidity during the day or year can affect the consistency of the results obtained in the printers. Thus, slight adjustments may be required to compensate for any variations if the environment is inconsistent (for example, to compensate for the cooler temperatures overnight).

Furthermore, the storage of the filaments has implications. For example, some locations usually require filament driers to compensate for high humidity, as the filaments tend to absorb water if possible, which decreases the printing quality.

Given the above, it is advisable to make some test prints with the materials of interest before printing with any new material. Depending on the case, such prints can be as easy as printing a small 3D model or more elaborate as printing a couple of models where some parameter is modified so that an optimal parameter can be tuned.

6.3.2.1   *3D Printing Materials and Their Implications*

Although the general printing parameters (extruder and bed temperature) for each material are usually described on a label attached to the roll of material, it is advisable to do research before printing, as some materials may require a particular type of nozzle or even the printer may require a special arrangement.

For example, let us suppose it is desired to print a 3D model with a metal or carbon fiber doping filament. In that case, it may be convenient to exchange the nozzle used during printing for one with a hardened material, since the metal or carbon fiber doping can be abrasive and wear the extruder bore, altering its size and damaging the extruder.

Moreover, some materials require systems that feed the extruder in a particular way. For example, suppose one desires to use a flexible filament. In that case, it must be "pulled" into the extruder instead of "pushed" into it. If one uses a printer that "pushes" it, there is a high risk of the filament getting stuck in the guide probe and stopping the extruder feeding. This can not only ruin the print but potentially damage the feeder motor or overheat the empty extruder.

Additionally, there are also some materials (e.g., ABS (acrylonitrile butadiene styrene) [1], ASA (acrylonitrile styrene acrylate) [2]) that, due to their properties, have some thermal contraction during printing [1, 2]. Hence, they tend to detach from the printing bed, have warping on flat surfaces, or retract the model corners. To solve this problem, this kind of material requires that the printer has a box, generating a controlled and warm environment that decreases these thermal contractions.

Lastly, some materials may release toxic gases during printing (e.g., ABS, Nylon) [3]. For these materials, it is usually convenient to have some filtering or ventilation system, ensuring that the people around the printings are not exposed to these gases.

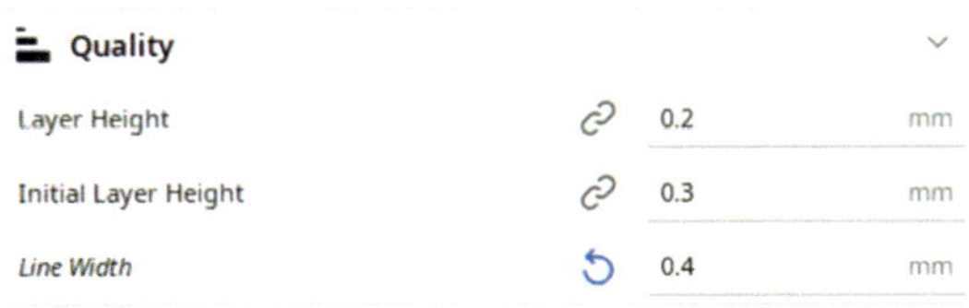

Fig. 6.6 Quality configuration panel

⊒ Quality			⌄
Layer Height	⌀	0.2	mm
Initial Layer Height	⌀	0.3	mm
Line Width	↺	0.4	mm

6.3.3   Printing Settings

The last section of Ultimaker Cura, contains a large set of parameters that alters how the 3D model is printed and how the material must be deposited. These parameters include but are not limited to, the material density distribution, the speeds of printing, or even the patterns that must be used to fill the volumes of the print.

In the case of Ultimaker Cura, each parameter usually has a drop-down window that, when hovering over the parameter in question, describes its implications and impact on other parameters in detail. However, to give an introductory description, the most relevant parameters of each section are described below.

6.3.3.1   *Quality*

The Quality panel (Fig. 6.6) allows the setting of several parameters related to the height of the layers as well as the width of each printed line. Among several implications, these parameters define the resolution used in the model's printing. Thus, one of two scenarios is generated: either thinner layers with better printing quality and longer printing times (as additional layers are required to complete the printing), or thicker layers that may not be attractive but decrease printing times.

Additionally, it is convenient to point out that the first layer of the print is usually an exception to the rule and has its own height. This exception promotes better adhesion of the 3D object to the printing bed, increasing the possibility of a better result.

6.3.3.2   *Walls*

The Walls panel (Fig. 6.7) allows the tuning of how many and how the walls should be made in the printout. Although trivial in appearance, this parameter generates a substantial change in the distribution of material within the object. An object with a larger number of walls tends to express slight deformations when a force is applied and has a higher moment of inertia, due to the material being deposited on the walls. However, when a deforming force exceeds the wall's strength, the model's structure becomes strongly compromised. Thus, this class of parts has poor elastic behaviors and thus becomes brittle at high stresses.

Given the above, the number of layers in the walls should be selected according to the application, with 2 walls for general-purpose parts and 4 to 6 walls for objects requiring stiffness.

Fig. 6.7 Walls configuration
panel

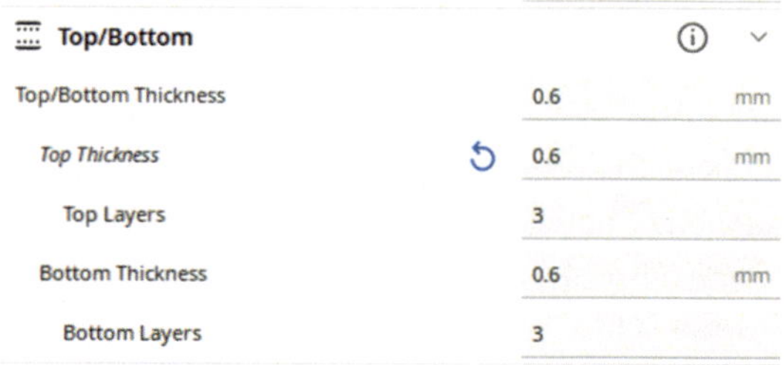

Walls   ⌄

Wall Thickness   0.8   mm

Wall Line Count   2

Horizontal Expansion   0.0   mm

Fig. 6.8 Top/Bottom
configuration panel

Top/Bottom   ⓘ ⌄

Top/Bottom Thickness   0.6   mm

Top Thickness   ↺   0.6   mm

Top Layers   3

Bottom Thickness   0.6   mm

Bottom Layers   3

6.3.3.3 *Top/Botton*

The Top/Botton section (Fig. 6.8) allows the configuration of a similar concept to the
Walls section but in the upper and lower regions of the model instead of the sides. On the
other hand, in some programs, it is possible to configure whether or not a smoothing is
desired at the top of the print (in such a smoothing, the extruder passes without depositing
material to allow the heat of the nozzle to improve the quality of the top layer). Nevertheless, this practice has now reduced its use thanks to the development of better printing
materials.

6.3.3.4 *Infill*

The infill panel (Fig. 6.9) is relevant, as it determines what percentage of the internal
volume of the object contains material and how much of it contains air.

Although one might initially infer that a high density is required to achieve structurally
strong parts, a density between 20 and 40% is usually more than sufficient for most
general applications. A density of 60%–70% is only required in isolated cases where
the part faces a more demanding structural load. And finally, densities of 90%–99% are
only required in exceptional cases where it is desired to take advantage of some material
quality.

An example of this last scenario can be found in the fabrication of the core of a 3D
printed motor, where it is desired to take advantage of the ferromagnetic qualities of a

Fig. 6.9 Infill configuration
panel

Infill   ⓘ ⌄

Infill Density   ↺   30.0   %

Infill Pattern   ↺ f_x Tri-Hexagon   ⌄

metal-doped filament. Therefore, the part is desired to have as much density as possible to increase the magnetic coupling of the components. Nevertheless, a comparative example of 3 different densities is presented in Fig. 6.10.

Lastly, this section determines the type of fill pattern used in the model. Among the most common patterns are the Tri-Hexagon and Gyroid patterns (Fig. 6.11). These patterns offer homogeneous material distribution and structural properties that allow the load to be distributed within the object. These properties have earned them a place in the 3D printing community. However, the Tri-Hexagon pattern is usually the most selected among these options. This preference is caused by the fact that it attenuates the generation of vibrations during the printing process concerning the Gyroid counterpart.

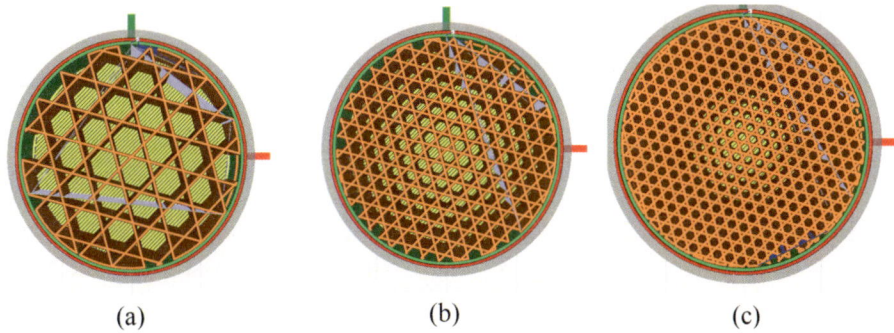

(a)   (b)   (c)

Fig. 6.10 Example of different densities in a 3D model. (a) 3D model with 30% density. (b) 3D model with 65% density. (c) 3D model with 97% density

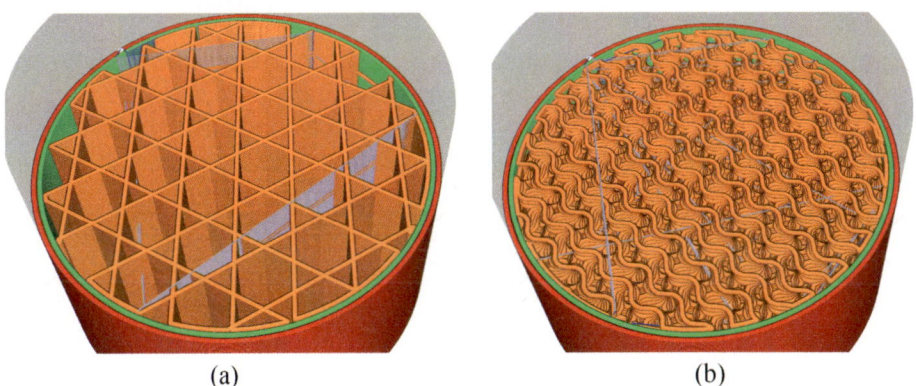

(a)   (b)

Fig. 6.11 Example of different infill patterns. (a) Tri-Hexagon infill pattern. (b) Gyroid infill pattern

Fig. 6.12 Material
configuration panel

6.3.3.5 *Material*

The Material Panel (Fig. 6.12) is a point of frequent checks every time the printer changes the material. This panel allows for altering the temperature of the extruder and the temperature of the printer bed, modulating the viscosity of the filament when extruding and altering the adhesion properties of the printer.

As previously indicated, these are some of the most critical parameters to tune for a successful print, so it is recommended to start with the values recommended by the manufacturers (usually employing a label on the filament roll) and adjust them to the printer environment through a set of calibration prints.

6.3.3.6 *Speed*

The Speed section (Fig. 6.13) allows the modulation of the speed of the extruder head during the printing of the 3D model in its different stages. This section, therefore, alters the speed of the first layers (usually slower to improve the adhesion to the bed), the speed of the fill (usually higher since imperfections in the fill are not visible), or the general speed of the walls, top and bottom of the model.

These parameters are defined by the properties and layout of the 3D printer and usually come preloaded in the printer profile. However, using some 3D benchmark models to calibrate and adjust these parameters to the most realistic values for the printer under test is usually a good practice.

Finally, it is convenient to reduce the printing speeds in 3D models that are tall and thin, as lower speeds tend to attenuate vibrations generated by the movements of the printed object and the extrusion head.

Fig. 6.13 Speed configuration
panel

Speed		
Print Speed	400.0	mm/s
Infill Speed	350.0	mm/s
Wall Speed	120.0	mm/s
Top/Bottom Speed	120.0	mm/s
Travel Speed	400.0	mm/s
Initial Layer Speed	25.0	mm/s

Fig. 6.14 Travel configuration panel

≝ Travel

Enable Retraction	✔
Retraction Distance	0.6	mm
Retraction Speed	40.0	mm/s
Retraction Minimum Travel	0.88	mm
Z Hop When Retracted	✔

6.3.3.7 *Travel*

The Travel section (Fig. 6.14) allows for the modulation of the speed movement required when the extrusion of an island in the model layer is finished and a movement over the air is required. Tuning these parameters is essential to control the behavior of the plastic when a stop in the extrusion is required. Among such parameters, the behavior of retractions is of great relevance. As its name implies, the retraction consists of retracting the filament to avoid generating a trace in the form of a thin thread while moving from one layer island to another. Adequate retraction allows the generation of clean prints that require little post-processing after printing.

In the travel panel, it is also possible to enable other procedures, such as the "Z Hop," which generates a slight jump in the extrusion head to help release the filament when retracting. An example of the implications of these parameters could be found when printing a 3D model of a hand with open fingers. Since the print is made layer by layer, the extruder has to "hop" from finger to finger. If the retractions are well calibrated, the model prints without problems. However, if it is not well calibrated, it is possible that at the end of the printing, some thread lines between the fingers appear as a residue of filament that was not cleanly released between each of the transitions.

6.3.3.8 *Cooling*

The Cooling panel (Fig. 6.15) modulates the fan speed of the printer and generates a minimum time to elapse between layers. These parameters give the print enough time to cool and solidify, which in turn promotes less deformation of the prints.

As a note to consider, over-cooling is usually preferable to under-cooling. Given the above, a high fan speed and sufficient minimum time between layers is desirable.

Fig. 6.15 Cooling configuration panel

❄ Cooling

Enable Print Cooling		✔
Fan Speed	f_x	100.0	%
Minimum Layer Time		5.0	s

Fig. 6.16 Support
configuration panel

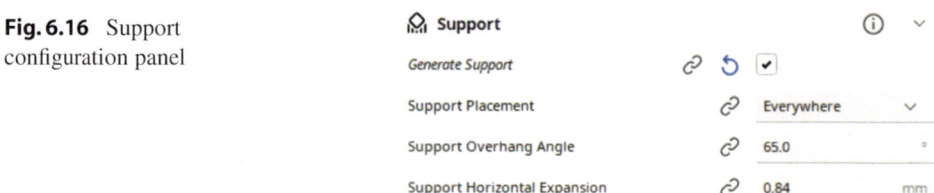

6.3.3.9 *Support*

The Support panel (Fig. 6.16) limits or completely removes the use of supports during the printing of the model. As previously mentioned, the supports are a set of automatically generated structures that act as a kind of scaffolding for the 3D model, allowing the printing of structures that, by their shape, would imply "printing in the air."

Usually, the support configuration includes a parameter to define one of 3 possible scenarios: (a) No supports are used, (b) Only supports that have direct contact with the print bed are used (avoiding supporting parts inside the 3D model), or (c) Supports are used in all locations that exceed a specified overhang angle.

Such specified overhang angle is a parameter that defines the threshold where supports should be considered (the angle after which the print is considered to be "floating" or "partially floating"). Although many 3D printer profiles already provide an appropriate overhang angle, there are also 3D Benchmark models that can be used to calibrate this parameter.

Finally, some slicing engines also consider the horizontal expansion of the supports separately (since the supports often have low density and have thermal properties outside the standard). Thus, this parameter enables a more appropriate spacing of the supports, making it easier to remove and reducing the marks on the 3D printed model.

6.3.3.10 *Build Plate Adhesion*

The Build Plate Adhesion section (Fig. 6.17) allows for the generation of structures that improve the adhesion of the part to the print bed. Although a well-calibrated printer does not usually require this kind of structure, some 3D models may have a mass distribution that requires using these structures to ensure that the model does not detach from the bed at a late printing stage.

For example, suppose one may want to print a tree model. In that case, there is a large mass in an elevated section of the model and a thin section to support and adhere. In such a scenario, splitting the 3D model to facilitate its printing or using an adhesion structure would be advisable.

Fig. 6.17 Build plate adhesion
panel

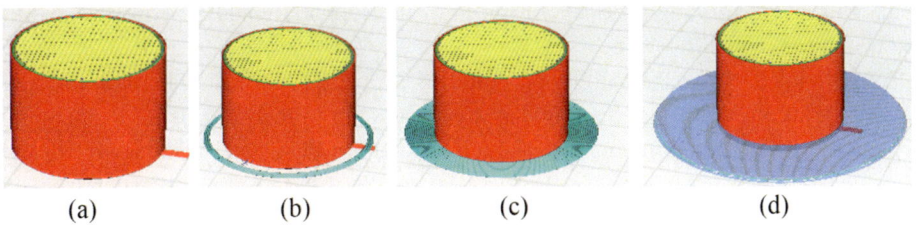

Fig. 6.18 Comparison of the various adhesion structures (a) None (b) Skirt (c) Brim (d) Raft

In Fig. 6.18, a comparison between the four types of structures traditionally used to improve adhesion to the printing bed is made. These structures allow a gradient of adhesion to be generated from simple adhesion to the use of complementary layers to secure the 3D model to the printing bed.

However, the use of these structures only improves the adhesion. However, it will not completely solve the deficiencies induced by a poor segmentation of the model, inherent deficiencies caused by a poor mass distribution, or printing orientation in the 3D model.

6.3.3.11 Dual Extrusion and Special Modes

Finally, it is possible to mention some sections that are potentially not required in many printers, such as dual extrusion (Fig. 6.19a) and special modes (Fig. 6.19b). These sections allow the use of multiple materials in a single print, or modify the order in which 3D objects are printed in case multiple 3D models are printed simultaneously.

The use of these sections is usually considered advanced, and some of these menus even have to be enabled through the preferences section of the slicing software, as they are not usually enabled by default.

However, by employing these sections (and a printer that allows their use), high-quality prints that feature multiple materials can be generated, allowing the intrinsic advantages of poly-material prints to be obtained. For example, one may wish to generate a complex structure that requires support in a space that physically would not allow them to be removed. In such a scenario, one could use a dual print where the part uses the material of interest (e.g., PLA (Polylactic Acid) [4]) in extruder one and PVA (Polyvinyl alcohol [5]) in extruder two.

Due to its properties, PVA is a water-soluble material [5], so its use in the previous example would allow the part to have all the necessary support during printing. However,

Fig. 6.19 Configuration Panels (a) Dual Extrusion (b) Special Modes

at the end of the printing process, all traces of the supports could be removed just by immersing the model in water.

Another interesting example can be found in the use of two materials with different thermal shrinkage constants. Such use makes it possible to quickly release the supports from prints using two different materials, for example, PLA and PETG (polyethylene terephthalate glycol) [6]. Thus, in these cases, it is possible to use one material in the printed object and the other in the supports, generating interaction points that tend to detach naturally due to the different thermal contractions.

Finally, as far as printing modes are concerned, this tool allows for the switch between parallel printing (where all objects are printed simultaneously) and sequential printing (where the first 3D model is printed and then the next one is started).

6.3.4   Model Printing

Once the 3D print has been configured, it is possible to perform the slicing and generate the printout file (.gcode file), which can be previewed in the PREVIEW section (available at the top of the screen). In the PREVIEW section, it is possible to use a set of sliders to analyze the different layers of the model and ensure that an instruction file with the desired properties is obtained (sections marked in Fig. 6.20).

Finally, it is possible to export the file to an SD or USB card (Fig. 6.20) for printing. This is usually a very straightforward and transparent process on current printers that may even calibrate the printing bed themselves, requiring only a couple of clicks after connecting the memory card to start printing. Nevertheless, checking the manufacturer's manual for the specific steps required for the interest 3D printer to be used is advisable.

6.3.5   Tips & Tricks for Successful Printings

To shorten the 3D printing learning process, this chapter closes with tips & tricks that can help improve 3D printing processes and increase the chances of success with this technology.

1. Make sure that there are no wind currents in your printing area.
2. Verify that the manufacturer's suggested print temperature settings for the target material is been used.
3. Ensure that the print bed has been leveled or that proper electronic calibration has been performed.
4. In case of adhesion problems to the print bed, use additional support structures and check the setting of the material. In the worst-case scenario, apply some washable adhesive (school adhesive sticks usually work, but it is better to check in a small

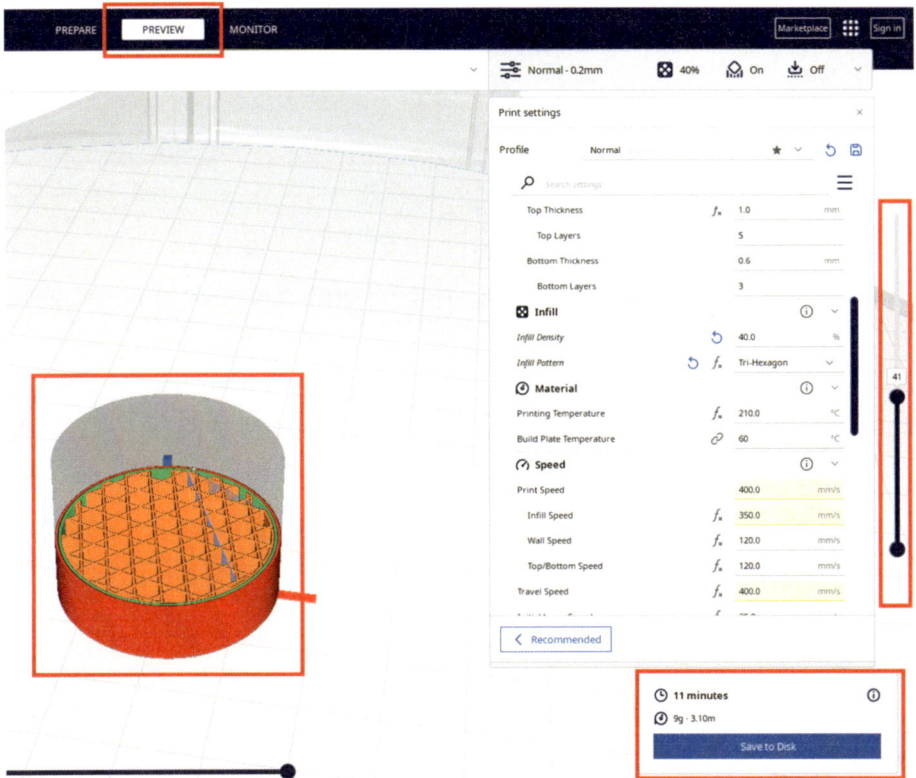

Fig. 6.20 Print preview screen with framed regions of interest

region of the print bed that the adhesive does not degrade due to temperature and damage the equipment).

5. Considering that the parts printed have a moment of inertia, an adjustment to the print speed as the height of the part increases may be required.

6. The best support is no support at all. Good designers try to design and orient their parts so that the use of supports is reduced to a minimum.

7. Two pieces may be better than one. Evaluate whether the part designed may have difficulties printing as a single unit, and if it is problematic, evaluate whether it is feasible to split the part into two separate units.

8. Good calibration can dramatically improve print performance. Although automatic profiles are usually very reliable, adjusting lateral expansions and considering design tolerances can improve the printing experience.

9. Remember that a 3D printer requires maintenance. Like any other machine, adjusting small details on the printer and following a preventive maintenance calendar can improve the performance of the equipment throughout its lifetime.

10. Remember that successful 3D printing starts at the design stage. Considering your processes from the design stage encourages subsequent stages to increase their likelihood of being successful and reduce manufacturing/prototyping time.

References

1. J. Ramian, J. Ramian, D. Dziob, Thermal deformations of thermoplast during 3D printing: Warping in the case of ABS. Materials, **14**(22), (Nov.2021). https://doi.org/10.3390/ma14227070
2. S. Guessasma, S. Belhabib, H. Nouri, Microstructure, thermal and mechanical behavior of 3D printed acrylonitrile styrene acrylate. Macromol Mater Eng, **304**(7), (Jul.2019). https://doi.org/10.1002/mame.201800793
3. S. Wojtyła, P. Klama, T. Baran, Is 3D printing safe? Analysis of the thermal treatment of thermoplastics: ABS, PLA, PET, and nylon. J. Occup. Environ. Hyg. **14**(6), D80–D85 (Jun.2017). https://doi.org/10.1080/15459624.2017.1285489
4. B.D.M. Matos et al., Evaluation of commercially available polylactic acid (PLA) filaments for 3D printing applications. J. Therm. Anal. Calorim. **137**(2), 555–562 (Jul.2019). https://doi.org/10.1007/s10973-018-7967-3
5. S. Mallakpour, F. Tabesh, C. M. Hussain, A new trend of using poly(vinyl alcohol) in 3D and 4D printing technologies: Process and applications. Adv. Colloid Interface Sci. **301** Elsevier B.V., (Mar. 01, 2022). https://doi.org/10.1016/j.cis.2022.102605
6. S. Valvez, A. P. Silva, P. N. B. Reis, Optimization of printing parameters to maximize the mechanical properties of 3D-Printed PETG-Based Parts. Polymers (Basel), **14**(13), (Jul.2022). https://doi.org/10.3390/polym14132564

Prototyping

7

7.1   Motor Prototyping

In this chapter the design process of various electric motors will be broken down part by part so that the reader will have a clear understanding of the process itself, since motor prototyping encompasses many fields of study in order to be successful.

The most important thing for the design of motors is to know their configuration, for example, the quantity, height and length of the poles of the magnets, as well as the teeth of the coils, if it is an axial or radial motor, internal rotor IR (IN-RUNNER) or external OR (OUT-RUNNER) and a long etcetera, it should be emphasized that it is important to properly calculate its parameters.

Another important aspect to design is to start with the part that cannot be manufactured, the best way to design is generate those parts in the 3D design software, in order to use these digitized bodies to compare with the sketch, speeding up the design process and eliminating any kind of guesswork error that the designer thinks about the measures that are given to the digital prototype against the real ones, with this process the software performs the calculations, and the user only needs to worry about how to design the motor. The prefabricated elements can be a lot, in this book could be the bearings, the magnets, even the shaft and as in the case of the last design, the teeth of the coils, in short, any part which is not manufactured by 3D printing, for the purposes of this book.

Knowing the configuration and the prefabricated parts, it is necessary to know the requirements of the motor, these are given by the configuration and the prefabricated elements, such as the length and width of the teeth, the separation of where the core ends with the cover, if it is OR that's necessary to put an outlet for the cables and as in the prefabricated parts a long etcetera. Understanding these concepts at the moment of design will make the process more enjoyable, as well as more efficient and entertaining the assembly of the motor.

© The Author(s), under exclusive license to Springer Nature Switzerland AG 2025   173
E. Cuevas et al., *DC Motors*, Synthesis Lectures on Engineering, Science, and
Technology, https://doi.org/10.1007/978-3-031-64354-5_7

Fig. 7.1 Motor 600w

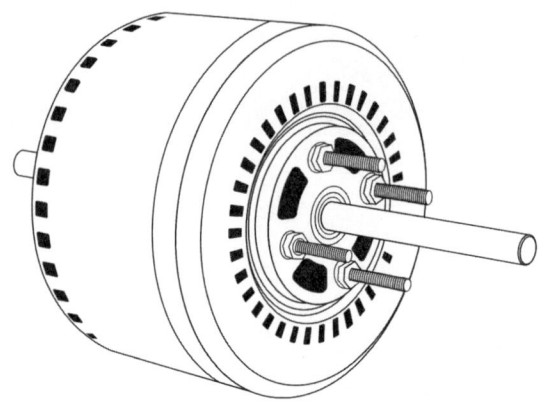

7.2   Inspiration

The design process of electromechanical machines has a technical part which requires adequate knowledge to understand the operation of the machine, however, creative knowledge is required to innovate and create something new, it is impossible to design something already existing due to the same definition of the word design: "Original concept of an object or work intended for mass production".

On the internet you can find prefabricated designs of certain motors, such as a very famous one using the HALBACH configuration of magnets [1], however, it is necessary to clarify that this type of plastic-based motors are less efficient than motors normally made with silicon steel cores due to their properties described in this book (Fig. 7.1).

Although there are certainly many designs that can be extracted from real electric bicycle motors, the performance will not be the same due to the characteristics of the materials in question (3D Filament), many of them influence the operation of the motor, but the most important to highlight is the magnetic permeability of the metal against the 3D printing plastic, in the market there are a lot of 3D printing filaments, However, not even the magnetic PLA that has iron shavings, managing to generate the greatest magnetic attraction (talking about 3D printing materials), manages to resemble the capabilities of the metal to focus the magnetic field (this can be reflected when approaching a magnet toward solid piece of printing and compare it with a solid piece of iron of the same dimensions).

7.3   Parts of the Motor

The kinematics of a motor generates rotary motion, this binds the motor to have a balance about the pivot point or also called motor's axis. The motor will be balanced if the center of masses coincides with the center of masses, this can be achieved by making the motor symmetrical in any straight line passing through the center of masses, i.e., by splitting the motor in the middle at any possible angle, two halves will be symmetrical with respect to each other talking about de radial plane.

The parts of the motor, although they may vary depending on the application, there will always be crucial parts for proper operation, such as the motor shaft, the core of the windings, bearings, the outer casing of the motor, and in this case, applying 3D printing concepts to the parts, the core of the magnets, also as optional parts, the bushings of the bearings, as well as the core of the coils and magnets, these depend on the design itself (Fig. 7.2).

7.3.1   Motor's Axis

The shaft is the base of every motor, in it, all components for the motors are mounted, such as OR, this shaft will be static and for IR motors will be mobile, to design a shaft we will have to take into account the application, the manufacturing method and the material to be used, for the purpose of this book the manufacturing method will be 3D printing, which restricts the material to be used or the characteristics with which it will have.

In case the application does not require a solid shaft to support considerable loads, such as a drone motor application. The shaft could be completely 3D printed, giving as an advantage a more complex design without having a higher cost of work due to the nature of 3D printing, this gives the designer a wider range to capture their ideas, with which you can generate different designs with more specific applications. Although 3D printing has certain disadvantages such as the limitation of material and certainly some physical forms due to its manufacturing process, it has the advantage that the difference in work of a complex part compared to a simple part is negligible, compared to other manufacturing methods, which, although they can generate the same designs in different materials, prototyping in other materials requires manufacturing methods with specific knowledge and specialized machinery, For example, Fig. 7.3 shows the plastic shaft designed for a drone motor, in it there are slots that go from side to side of the shaft, for the 3D printer will be like making another piece, however, to do it in another material will be up to double or triple the work, there will even be occasions that the piece will be impossible to do without specialized machinery.

Due to the ease of manufacturing with which 3D printing has, it is easy to adapt to the preset forms, an example would be the Fig. 7.4 where a 2-inch aluminum tube is used to hold the bearing bushings and the core of the winding, as you can see the thickness will be given by the loads that are presented in the application, Generally in the materials industry the material is sold by weight, generating that a 2 inch tube of 1/8 of thickness costs about 10 times less than a solid bar of 2 inches of aluminum, this way, you can buy the right piece for the motor and not waste in unnecessary expenses.

Fig. 7.2 Motor's parts

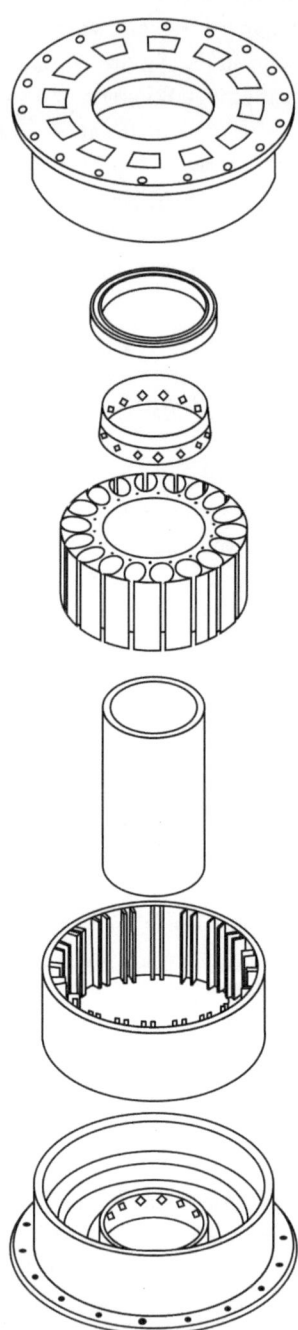

Fig. 7.3 Plastic shaft

Fig. 7.4 Metal shaft

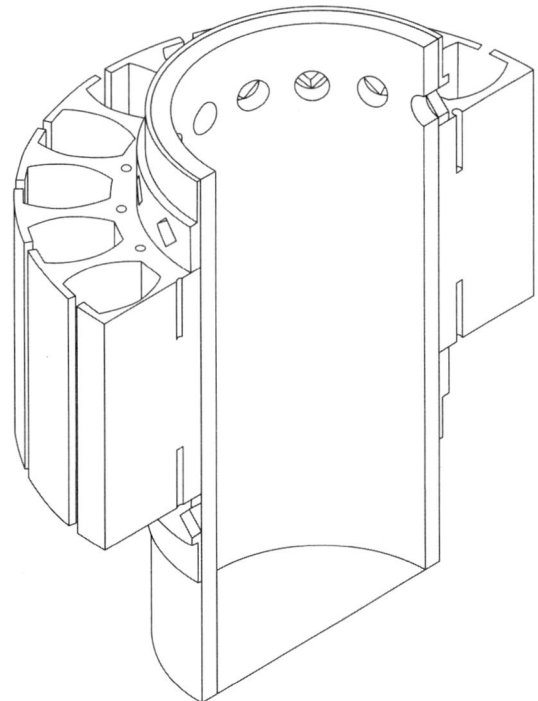

7.3.2   Core Winding

The core winding must be created to hold the coils in the right place, that is, aligned with the magnets and as close as possible without touching either the magnets or the magnet bases, depending on how it is designed.

The core winding is tied to the magnet configuration, whether it rotates the shaft (IR) or the housing (OR), the power of the motor, because the winding must have clearance to enter and whether it is OR the amount of space available inside the minimum diameter of the magnets.

7.3.3   Magnet Core

Like the winding, the magnets can have their respective base that assembles with the motor housing. In conventional brushless motors, the poles of the magnet are aligned to the center of the motor, either north or south, as shown in Fig. 7.5.

In addition to the normal distribution, a Halbach distribution can be used, this distribution consists of using a pair of secondary magnets next to the main magnet [2], achieving to focus the field of the main magnet in order to have more torque.

This configuration is achieved by placing the pole of the secondary magnet pointing to the main magnet, as shown in Fig. 7.6, where the north pole of main magnet is pointing to the center of the motor and the north poles of the secondary magnets are pointing to the main magnet.

Fig. 7.5 Normal arrangement of magnets

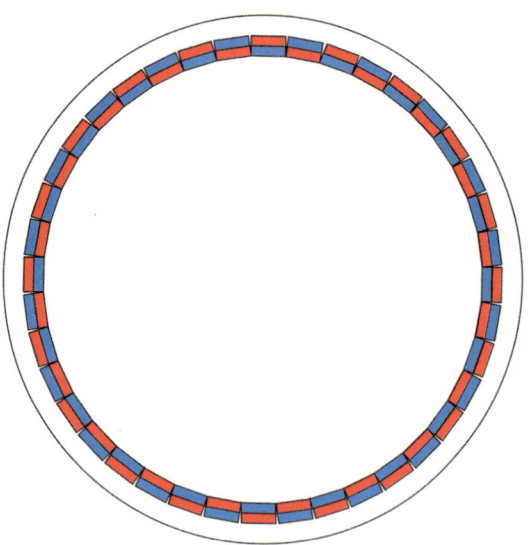

Fig. 7.6 Arrangement halbach

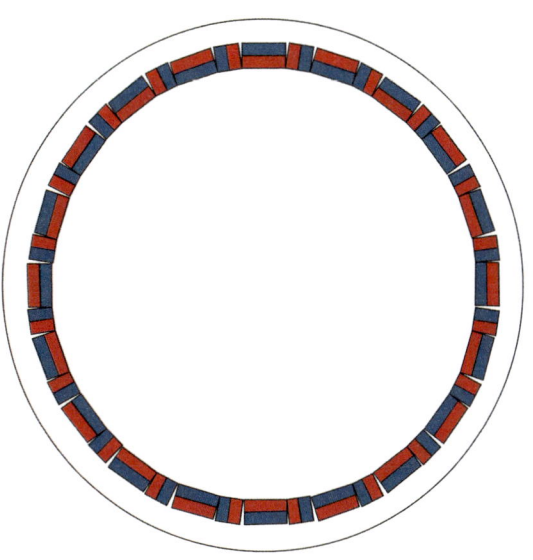

The normal distribution of magnets can be transformed into a distribution of fewer poles with the same magnets by using several magnets with the same pole one after the other. By having fewer poles the distance between coil teeth is greater, therefore, the motor can have more turns per tooth or you could wind the same amount of turns with a larger amount of wire, although you can certainly compensate the size of the coils with the quantity in the opposite case, the practicality of winding fewer teeth is greater. The difference of using several magnets in a row instead of one magnet, is because of the shape of the magnet, when the motor is running the distance between the coil tooth and the magnet will be more uniform in contrast to a straight magnet where the central part will be closer to the core and the edges will be farther away, It should be emphasized that a curved magnet would be even better in this aspect, however, the improvement would not be as noticeable and the acquisition of these magnets will depend on the diameter of each motor, not being able to be used in other diameters (Fig. 7.7).

Although the halbach configuration can certainly be reduced in a similar way, it would have to be done by using the secondary magnets at the end of each pole, as shown in Fig. 7.8.

7.3.4   Bearings

The movement of the prototype will be rotational; therefore, it is necessary to know the different types of ring bearings, which have the function of generating the least resistance to rotational movement. The bearings, although it is a fundamental part of most industrial and domestic machines, there is no manual to generalize all existing ones, however, each

Fig. 7.7 Arrangement four
magnets per pole

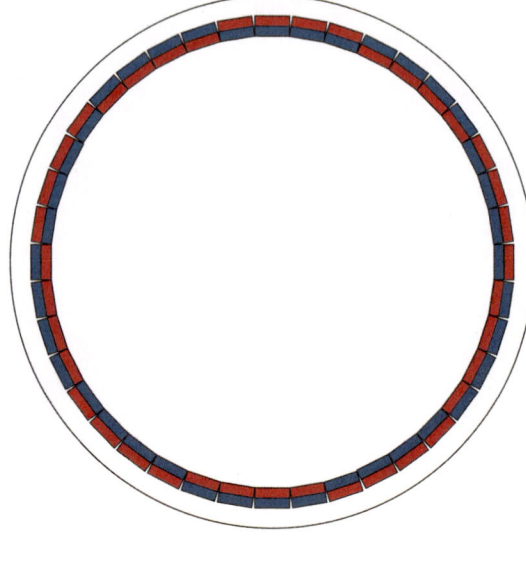

Fig. 7.8 Arrangement of four
magnets per pole in halbach

company has its catalog, with specifications to find the bearing that best offer for the
specific application, and even some companies have a series of steps to follow, such as
SKF, in its website is detailed a process to make more efficient the selection of bearings.
The first step is to know how the bearing will work, taking into account the forces that
interact with it, the life required for that application, the speed and acceleration to which
it will be subjected, however, the most important values to consider, would be, the space

included for the bearing in the design, measures such as axial or radial space required, the second step is to know the type of bearing that is needed, each has characteristics that make it better for certain applications according to their profile, this can be seen by the cross section of the bearing.

Figure 7.9 shows some types of bearings offered by SKF, as well as their interior.

As can be seen in Fig. 7.9 there are many types of bearings, including some are two or three encapsulated in one to reduce the disadvantages it has by itself, usually for motors made by 3D printing ball bearings (item a) can be used as a general purpose, because the mechanical properties of the filament are not very good compared to metal, however, in certain structures the forces can be distributed evenly over the entire plastic structure,

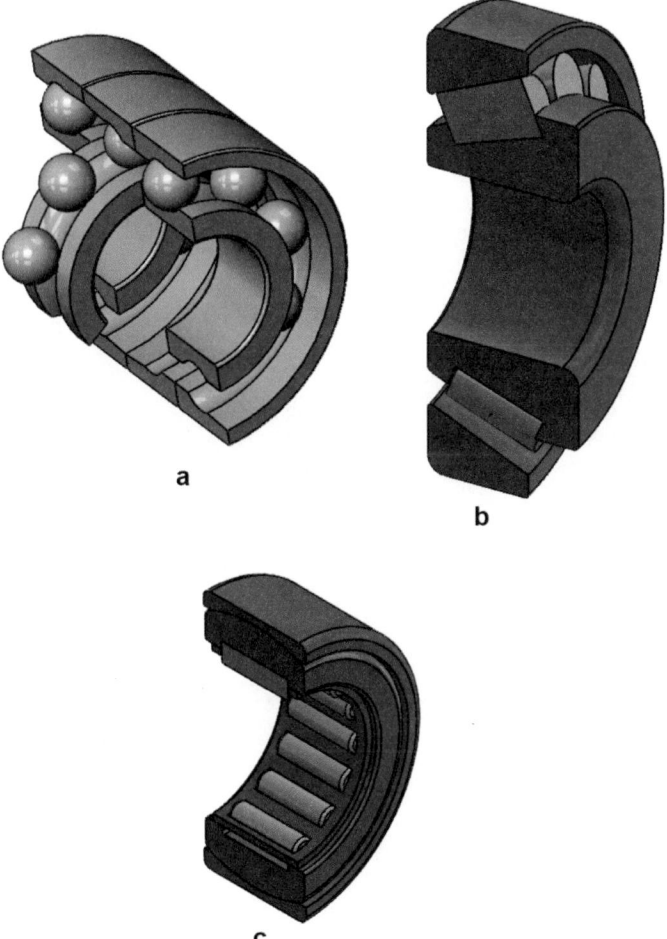

Fig. 7.9 Types of SKF bearings

thus achieving better mechanical properties and being able to use roller bearings (item B) which are used for larger loads than ball bearings due to their better support over the entire roller surface, for small motors that need to withstand large loads, needle roller bearings are ideal (item C).

7.3.4.1 Selection of Bearings

The selection of a bearing requires some technical expertise about them, and the idea is to select the theoretical number of working hours that the bearing can withstand without failure. This book will concentrate on the use of ball bearings, in the SKF manual [3] on page 322 begin the ball bearings they offer, in those tables are the outer and inner diameters, their designation, the basic load capacity, as well as the fatigue capacity, the nominal speeds as the limiting speed, its mass and the code of each bearing. The selection process is an iterative process where it starts with a specific characteristic of the bearing, such as any of the above mentioned.

Features:

- Inner diameter = 50
- RPM = 2000
- Work temperture = 60°C

For this example we will start with an inner diameter of 50 mm, in the corresponding table there are 7 types of bearings, so we will start with the outer diameter of 80 mm and 16 mm in height which will be the bearing of average measures with the designation of 6010, so you can make an estimate if the next bearing to be calculated has to be larger or smaller thus reducing the amount of calculations needed to reach the optimal bearing.

In the Table 7.1 the values from the data sheet published by SKF [3] are represented considering the bearing codes in the last column k. For the other values d corresponds to the inner diameter, as well as D to the outer diameter, B will be the bearing thickness, C is the dynamic load capacity of the bearing, as well as C_0 is the static one, P_u corresponds to the fatigue limit, V_r and V_l will be the reference and limit speeds correspondingly and finally M is the mass of the bearing itself.

Table 7.1 Sizes of bearings

d	D	B	C	C_0	P_u	V_r	V_l	M	k
50	65	7	6.76	6.8	0.285	20000	13000	0.052	61810
	72	12	14.6	11.8	0.5	19000	12000	0.14	61910
	80	10	16.8	11.4	0.56	18000	11000	0.18	*16010
	80	16	22.9	16	0.71	18000	11000	0.26	*6010

The second step is to select the lubrication, this is done by obtaining d_m, from the following formula:

$$d_m = 0.5(d + D)$$

where d is the inside diameter and D is the outside diameter. For this example, we obtain:

$$d_m = 0.5(d + D) = 0.5(50 + 80) = 65$$

Obtaining the constant d_m and assuming that it will work at a speed of 2000 RPM, you can clear the nominal viscosity on the graph [3] of Fig. 7.10, which represents a logarithmic scale where the horizontal axis represents the constant d_m, the vertical axis represents the nominal viscosity and the lines diagonal represent the RPM at which the bearing will work, to obtain the viscosity, find the viscosity point that is horizontally aligned with the point on the 2000 RPM line that is vertically aligned with the d_m value of 65, therefore the nominal viscosity value will be 12 $\frac{mm^2}{s}$ approximately.

After obtaining the viscosity, the type of lubricant to be used in the bearing is determined. Figure 7.11 shows the graph [3] of the lubricant where the horizontal axis represents the working temperature while the vertical axis represents the nominal viscosity; the lubricant to be used will be the logarithmic line at the intersection of the vertical line that cuts the temperature with the horizontal line that cuts the nominal viscosity.

With these values we can obtain the value of the viscosity ratio, given by the following formula:

$$k = \frac{V_R}{V_N}$$

where V_R is the actual viscosity value and V_N is the nominal viscosity value, replace:

$$k = \frac{12}{22} = 0.54$$

The next step would be to obtain the contamination factor [3] from Table 7.2, where the contamination factor depends on the constant d_m calculated above, if it is less than 100 the left column will be taken otherwise the right column will be taken. Contamination is subjective, but there are general features to determine which contamination will be the most adequate.

Extreme cleanliness is when the contamination particle size is approximately the thickness of the lubricant film, this usually only occurs in laboratory conditions. High cleanliness is generated when the lubricating oil has a very fine filtration, this corresponds to bearings with lifetime lubricated shields. Light contamination occurs with the entry of wear particles into the bearings and slight ingress of contaminants. Typical contamination occurs when there is a considerable amount of contaminating particles, compromising the

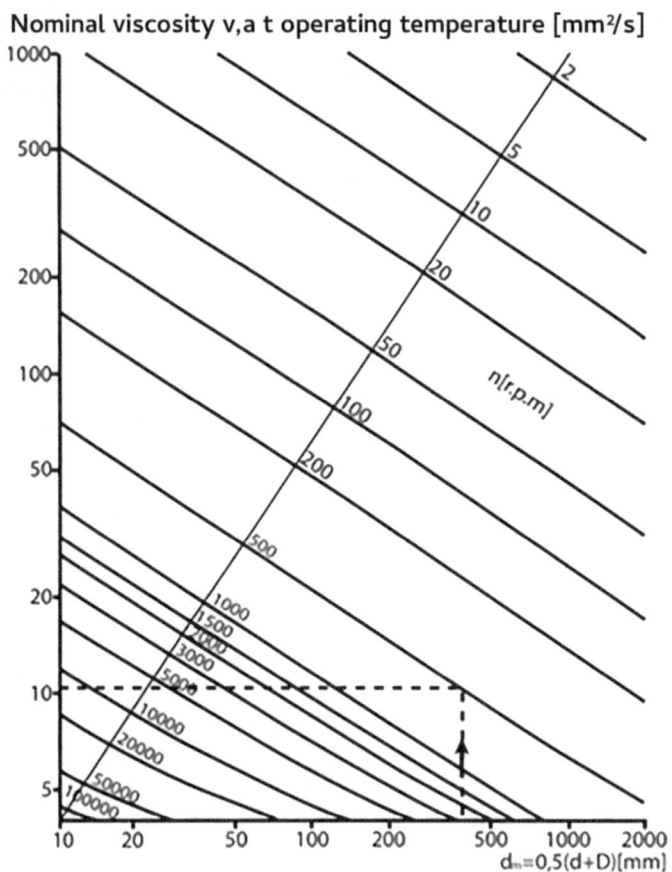

Fig. 7.10 Graph of viscosity

useful life of the lubricant. Severe contamination occurs when there is no effective sealing of the bearing and extreme contamination is when the contamination factor is off the scale impossible to calculate.

The valor a_{skf} value is a constant that helps to determine more accurately the bearing life, this is obtained by first obtaining the value of the following equation:

$$v = n_c \frac{P_u}{P}$$

where P_u is the fatigue limit load, obtained from the tables, P is the bearing load and n_c is the contamination factor, substituting:

$$v = 0.2 \frac{0.71 kN}{0.98 kN} = 0.144$$

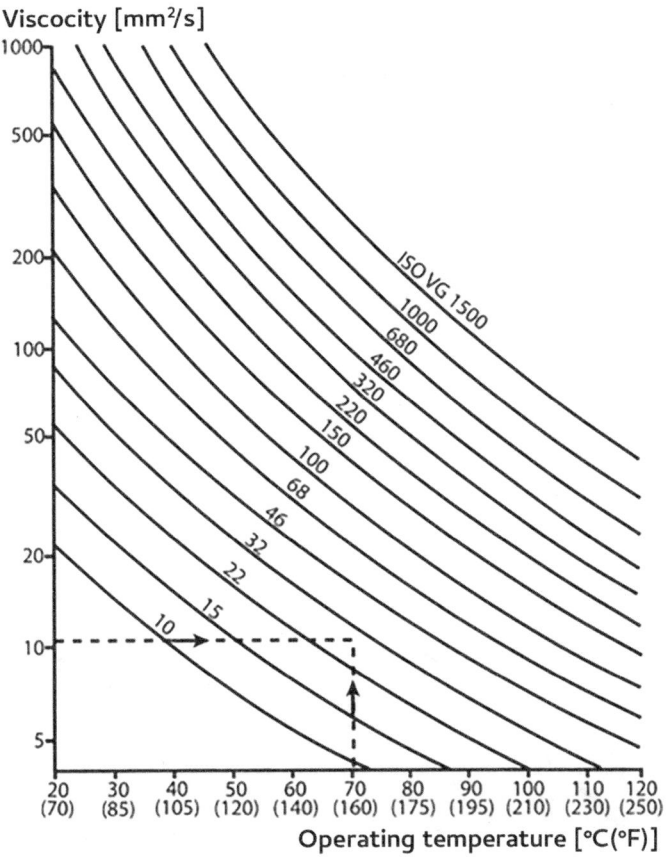

Fig. 7.11 Graph of lubricant

Table 7.2 Contamination factor

Condition	Factor n_c	
	$d_m < 100\,\mathrm{mm}$	$d_m \geq 100\,\mathrm{mm}$
Extreme cleanliness	1	1
High Cleanliness	0.8–0.6	0.9–0.8
Normal Cleanliness	0.6–0.5	0.8–0.6
Light contamination	0.5–0.3	0.6–0.4
Typical contamination	0.3–0.1	0.4–0.2
Severe contamination	0.1–0	0.1–0
Extreme pollution	0	0

Once the value is obtained, the value of the constant a_{skf} can be obtained by means of the graph [3] of Fig. 7.12, the obtained value will be the horizontal axis, a vertical line is placed on the obtained point and in the intersection on the line of the viscosity relation previously calculated a horizontal line is created, the value of cut of this line in the vertical axis will be the value of the constant a_{skf}, obtaining a value of 0.75 approximately.

To obtain the basic rated life of the ball bearing in revolutions, it must be calculated based on the following equation:

$$L_{10} = \left(\frac{C}{P}\right)^3$$

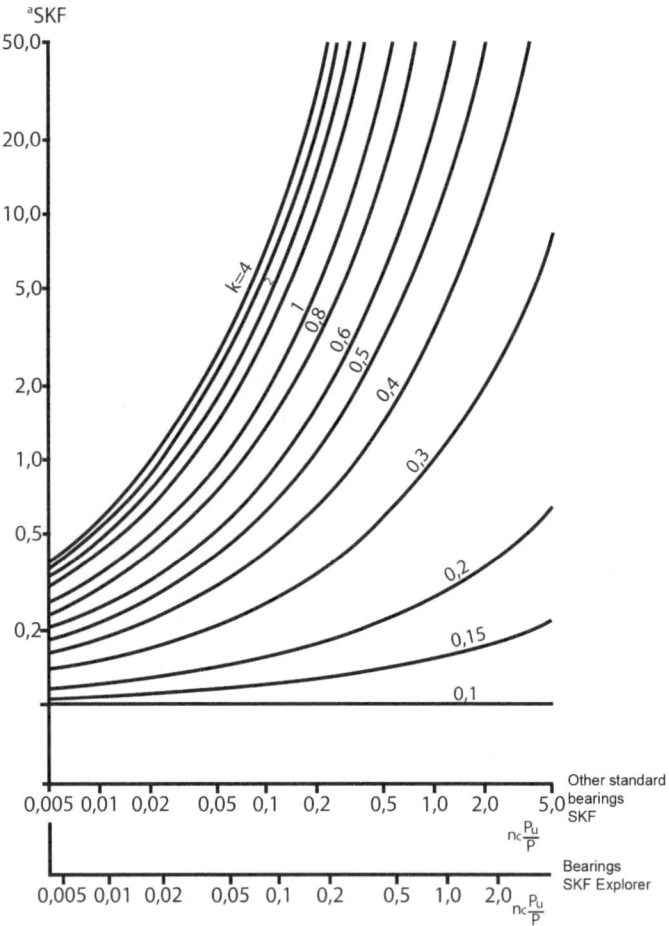

Fig. 7.12 Graph of SKF constant

where C is the load capacity and is obtained from the bearing table, and P is the load applied to the bearing, substituting the value of C obtained from the tables and the value of P that was stated in the initial conditions we obtain:

$$L_{10} = \left(\frac{22.9kN}{0.98kN}\right)^3 = 12759.34 \text{ millon of revolutions}$$

The following formula is used to convert them to working hours:

$$L_{10h} = \frac{10^6}{60n}L_{10}$$

where n is the RPM at which the engine will be running, substituting:

$$L_{10h} = \frac{10^6}{60(2000)}12759.34 = 106327.8333$$

To obtain the SKF rated life, the following equation is used:

$$L_m = a_1 a_{skf} L_{10h}$$

where L_m is the nominal useful life of SKF, a_1 is the factor of reliability of SKF, a_{skf} is the factor SKF y L_{10h} is the basic useful life in hours, replacing:

$$L_m = 1(0.75)(106327.8333) = 79745.87498$$

In addition to the bearing life it is also important to know other parameters such as the minimum working load, when the bearing does not have a load strong enough to press it, the bearing can fail by skidding, this causes both the internal part of the bearing and the external part to rotate simultaneously, wearing the parts together with the bearing rings or the rings themselves, to know the minimum force required by a bearing is obtained with the following formula:

$$F_{rm} = k_r\left(6 + \frac{4n}{n_r}\right)\left(\frac{d_m}{100}\right)^2$$

where F_{rm} is the minimal load, k_r y n_r are calculation factors obtained from the SKF website, n is the speed at which the bearing will operate, d_m the average of the internal and external diameter.

7.3.5 Bushings

There are parts of the motors that do not always fit together, this generates unwanted clearances between parts of the motor, the bushings have the function of filling the empty

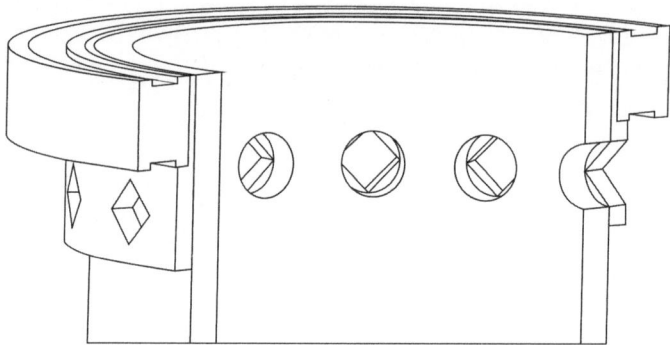

Fig. 7.13 Bushing 2-inch

space that is required with material reducing the clearances, usually when the manufacturing processes of a part are not precise or for adaptation of prefabricated parts, such as
an aluminum tube with respect to a bearing, as can be seen in Fig. 7.13. Where a 50 mm
internal diameter bearing is assembled with a 2-inch diameter tube, the tube size is not
precise enough as required, so a smaller one is acquired to later design a bushing with
the exact measurements to center the bearing well.

In Fig. 7.13 you can see a bushing, this separates both the bearing from the tube radially, and the bearing from the winding core axially, it should be noted that the bushings are
designed to have measures with considerable precision to maintain the smallest possible
clearance.

7.3.6 Motor's Housing

The outer shell of the motor is called housing, this has two main functions: cover the
motor from external agents and give rigidity to the structure. The structure to place the
magnets as close as possible to the motor coils. In addition, in case that the motor is
going to mounted in bicycle the housing needs a coupling for the bicycle ring, obtaining
another functionality (Fig. 7.14).

The motor must have two parts to transform electrical energy into mechanical energy,
a static part and a moving part, this will depend on the configuration of the motor, more
specifically if it is OR when the housing is mobile, or IR when the housing is static, and
in turn the configuration will depend on the application, as it can be a DC electric motor
(item A) adapting to the IR configuration or a hub motor (item B) for bicycle taking an
OR configuration.

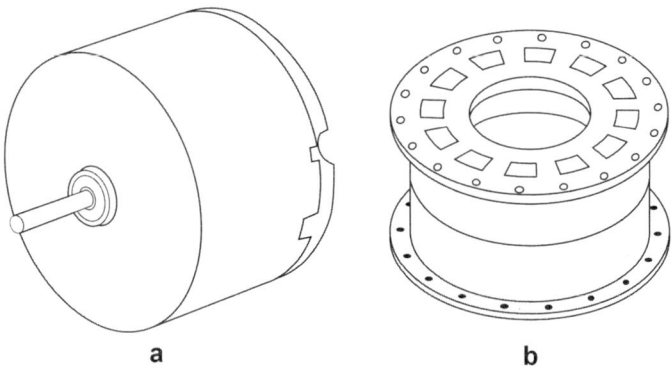

a   b

Fig. 7.14 Types of housings

7.4   Sketch

The sketch abstracts of most of the dimensions of a 3D part in 2D drawing, a projection of the 3D figure to a 2D plane and because dimensions will be lost by the very nature of the transformation is always better to use the plane that contains as many dimensions as possible, for example, a cylinder can be represented in 3 measures, In order to generate the part in any design software it is necessary to generate a sketch that contains the dimensions of the inner and outer diameter since this is the sketch that has the largest number of dimensions to then apply a tool such as extrusion. Figure 7.15 shows the necessary dimensions to generate a cylinder.

As in the previous example, motor designs can start from a basic sketch that varies for each motor. To make the design more enjoyable and interesting, it is necessary to know the position relationships and how they interact with the other dimensions, in the design, there are parts that are impossible to make, for example, having two symmetrical points and wanting to put a different distance to the line of symmetry to each point is impossible to generate by the same concept of symmetry. To design a motor, you need to separate in a circular arrangement the magnets. Although in most design software there are 2D tools that facilitate the same design, it is better to know the basics of trigonometry of the sketch, in order to have more flexibility when designing.

The first measure to know in the design is the minimum circumference that the magnets can have when placed in a circular way when the magnet is square, in case of having a curved magnet the circumference is limited to the radius of the arc of the magnet. Starting from the minimum circumference the rest of the motor can be generated, unless of course there are other restrictions.

This can be calculated by defining the number of poles to be used in the motor, once the number of poles is obtained a basic sketch can be generated with position relations

Fig. 7.15 Cylinder

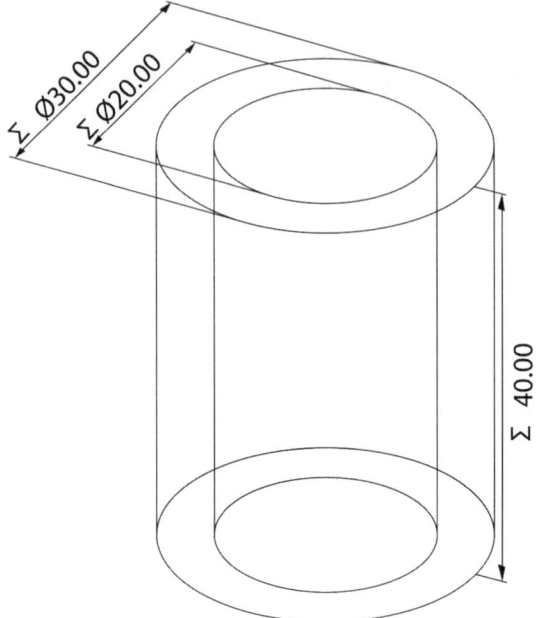

and the same software will adjust the geometry, however, trigonometry can give us a clearer idea of the physically possible designs, the sketch is represented in Fig. 7.16.

In the previous sketch we can observe a distance D which corresponds to the width of the magnet, the distance T which corresponds to the distance that separates in a straight

Fig. 7.16 Sketch of analysis

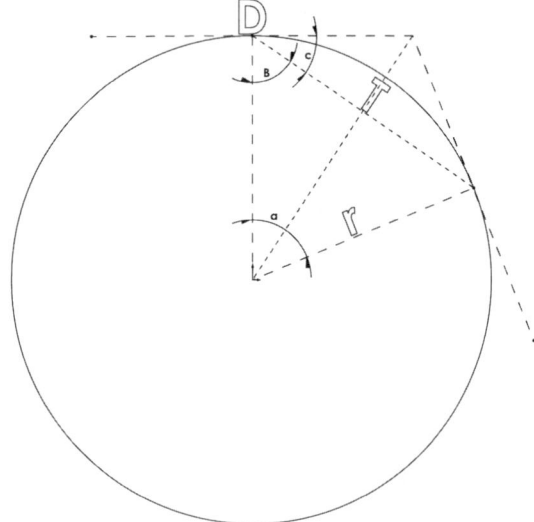

line a magnet from the next one in the arrangement, the angle a corresponding to the angular separation of the lines that go from the center of the circumference to the center of each magnet, the angle B which corresponds to the angle of the line r with the line T and finally the angle between the line D and T labeled as C.

The variables that we have would be the number and width of the magnets, the variable a can be obtained with the following formula:

$$a = \frac{360}{n_i}$$

where a is the angle described above and n_i is the number of magnets to use, this formula is cleared from the number of degrees that has a circle divided by the number of magnets.

Once obtained the angle a it is possible to clear B by the property of the summation of the angles of a triangle, which tells us that the sum of the angles of a triangle is 180 degrees, resulting in the following formula:

$$180 = 2B + a$$

where B is the angel between T y R, and it can obtained B by the next way:

$$B = \frac{180 - a}{2}$$

When obtaining B we can clear C being this the Angle between D and T, and can be cleared because the line D is tangent to the circumference and therefore perpendicular to the radius R, resulting in the following formula:

$$90 = C + B$$

where C can be clear:

$$C = 90 - B$$

From angle C the distance T can be obtained by applying the cosine law for right triangles, which will be the one formed by lines D and T and because line T cuts D in half, half of each line is taken, therefore, the hypotenuse will be half of D and the adjacent leg half of T resulting in the following formula:

$$\cos c = \frac{\frac{T}{2}}{\frac{D}{2}} = \frac{T}{D}$$

obtained the variable T it is possible clear R from the same equation applied to B angle where R will be hypotenuse:

$$\cos B = \frac{\frac{T}{2}}{R}$$

clearing R:

$$R = \frac{T}{2 \cos B}$$

where R is obtenied the minimum diameter of placing a certain number of magnets in a circular arrangement can be calculated by multiplying by two:

$$D = 2 * R$$

The above procedure can help not only in obtaining the diameter but also in the visualization of the dimensions and mates, how they interact with each other to make the design simpler.

Figure 7.17 has the complete sketch and it can be observed that only with two dimensions it is possible to define the diameter of the minimum circumference, this is achieved due to the position relations.

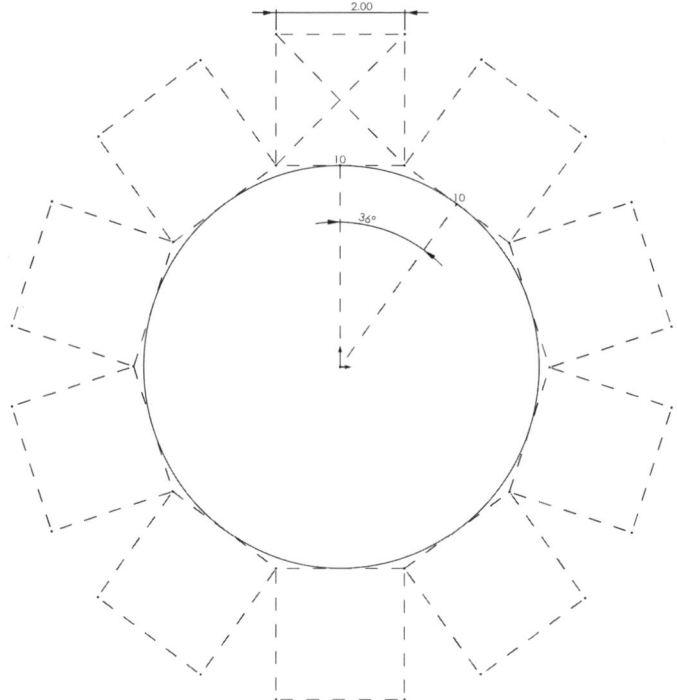

Fig. 7.17 Sketch of magnets

The starting sketch for the design of any motor in this book is shown in Fig. 7.17, which shows the arrangement of the magnets, and a circle which represents the smallest diameter of an array of 10 magnets.

7.5 Prototype One (18-20)

The first prototype of this book is inspired by a configuration widely used in HUB motors, to be precise the 18-20 configuration of coils and magnets respectively and was designed as a prototype for performance testing.

After deciding that the motor will be OR, will have a configuration 18-20 coils and magnets respectively and will be designed based on magnets with measures $5 \times 10 \times 40$ N35, it is possible to generate a sketch. The first thing that needs to be done is to start with parts that are already manufactured and especially selected for the design, as in this case are the magnets.

Figure 7.18 represents the exploded view of the motor, in this view are all the parts of the motor as if it were disassembled, in addition, it is an intuitive way to see how the assembly will be done with real parts.

7.5.1 Magnet's Base

Figure 7.19 shows the motor of prototype 1 split in half while the parts are arranged so that they do not interfere with each other, in the model are the main parts, these are the ones that accommodate the coils and magnets to generate a good interaction between these same elements.

Figure 7.20 shows a detailed view of the magnet bases. These bases were designed to use the circular die operation to overlap each other, so that a base with two protrusions can be extruded to hold the magnets on both sides and also reduce the space between magnets while increasing the efficiency of the motor, that is why the angle dimension is taken from the inner line of the left base to the outer line of the right base.

Figure 7.21 (A) shows as a drawing the base of the magnets with its inner and outer radius of the circumference of the magnets, as well as the outer diameter of the cylinder that provides rigidity to the piece, in addition, in (B) is the same piece in isometric view, where you can see its height.

7.5.2 Core

When designing the core, the minimum diameter is known, for the magnet arrangement together with their respective bases, with this diameter the motor core can be designed.

Fig. 7.18 Prototype one,
exploded view

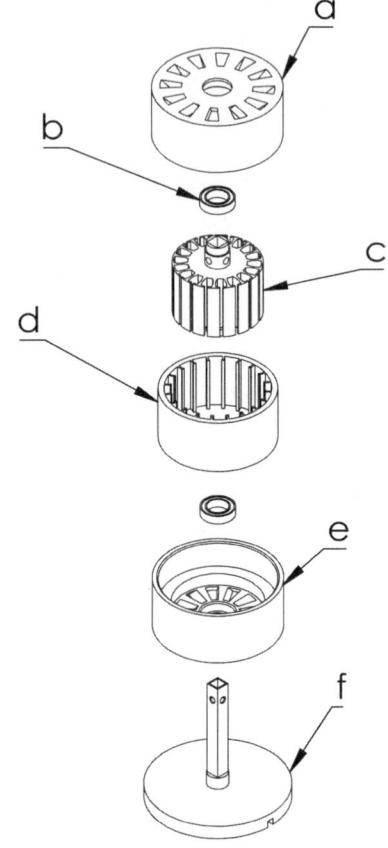

Figure 7.23 shows a coil tooth and the minimum diameter of the magnet array extracted from the previous design. It can be observed the coil tooth, designed to apply the circular matrix operation after, this was designed to have the largest amount of space in the center in order to put the largest number of turns, and thus improve the power-space ratio, it can be seen how the joints with the core and the tongue have a larger contact surface to make the piece more rigid, as well as more resistant.

The motors need to be unbalanced by nature, where the ratio between the magnets and the coils has a number as close as possible to one, however, for all types of three-phase brushless motor is necessary that the number of coils is a number that can be divided by three results in a whole number, so that the motor coils are balanced and the motor can rotate in a harmonious way. For this reason, the number 18 of teeth was chosen giving as a result six to the quotient of three and being the closest number to 20 below the same since the space is more reduced to smaller diameter, this being the case to be inside the core of the magnets and have sufficient clearance for any slack that may have the motor.

Fig. 7.19 Prototype one,
assembly with accommodated
section view

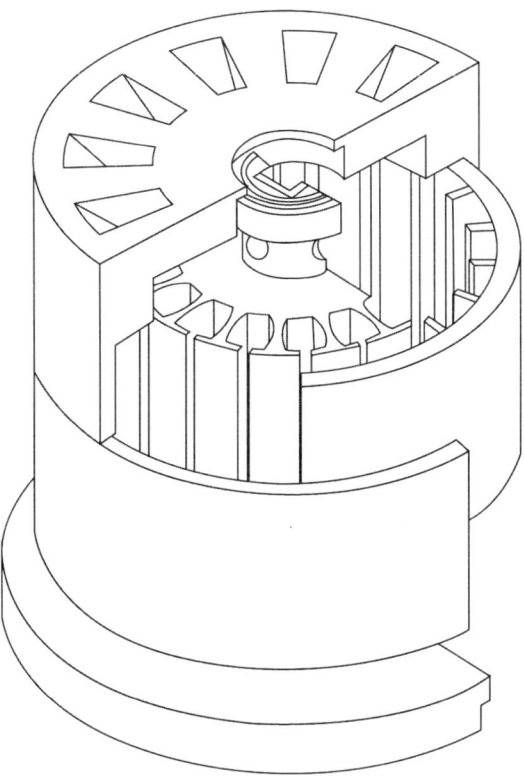

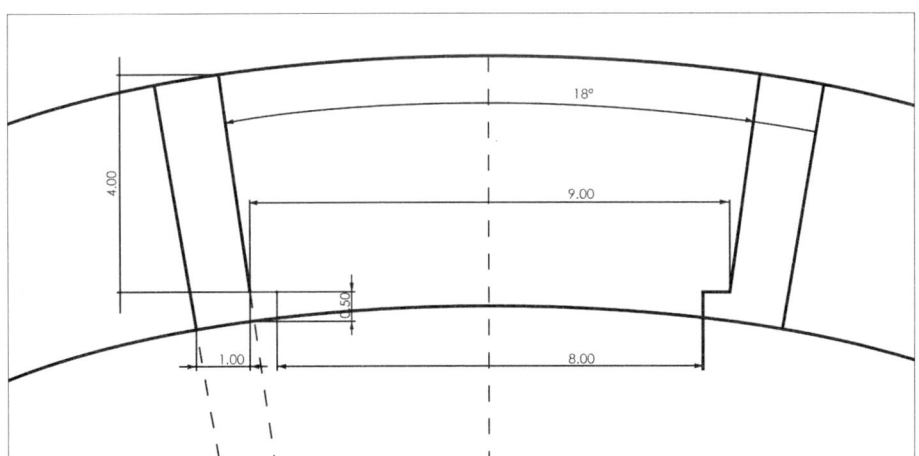

Fig. 7.20 Prototype one, sketch magnet's base

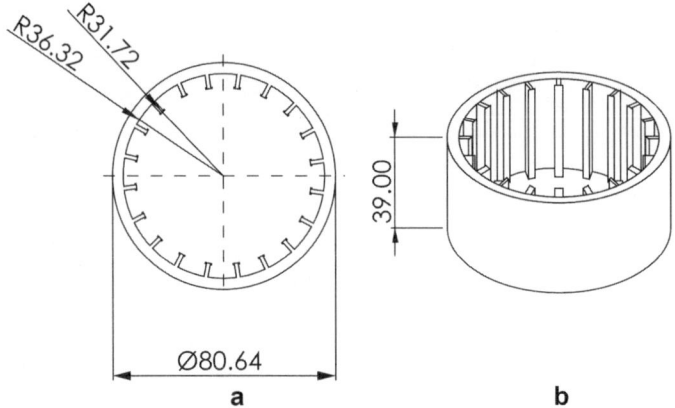

Fig. 7.21 Prototype one, magnets' base

Figure 7.23 illustrates the design of the core of the coils, which will hold the coils in the correct position to interact with the magnets, as well as is focusing the magnetic field radially pointing to the center of the motor. In addition, the figure shows the measurements of this part commonly called dimensions, if we observe in more detail we can see that these dimensions are defined from the sketch of the core of the coils in Fig. 7.22 and the height is given by the height of the magnet, if we want to build a more efficient motor it is necessary to cover the largest possible magnetic interaction. On the other hand, it is observed that this core has two extrusions above and below, which are symmetrical in the piece, these two protuberances are separators integrated in the same model, precisely to separate the bearing and that at the moment of designing the casing this one does not get to rub the windings, this separation is given by the first cylinder, since the second cylinder will be to center the bearing, in the first extrusion there are 4 holes through which the winding cables will pass, this is because it is necessary to use the shaft as a conduit, therefore a square cut was generated across the entire shaft, the nature of the OR motors makes this the only static place that has contact with the outside of the motor.

7.5.3   Lid

The housing for the design of 3D printed brushless motors is for OR one, is the last piece to design, talking about the parts of the motor, this must have certain functionalities, the main one would be to center in position the base of the magnets, this is achieved by two main restrictions, one is the radial centering, where this design uses the outer layer F to restrict the base, on the other hand, the axial centering is managed by a protrusion D that restricts the movement up and down. In addition to holding the base with respect to this part, this same part has to be axially and radially centered also on the core, this was

Fig. 7.22 Prototype one, sketch of tooth

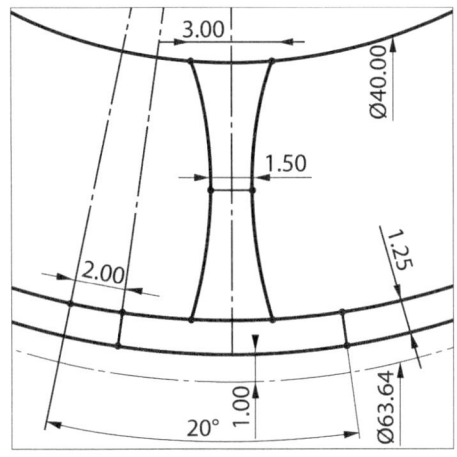

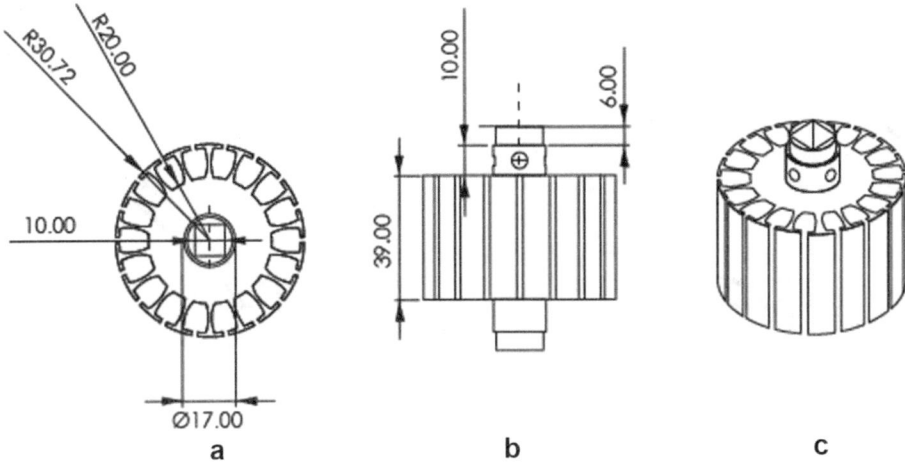

Fig. 7.23 Prototype one, coils' core

achieved by means of the bore E that manages to place the bearing in the radial center and restrict it axially as far as the bearing reaches, by joining the two halves it is completely axially restricted (Fig. 7.24).

7.5.4 Base of the Motor

The design specifications of the first prototype are to create a test prototype, therefore, a base was created from which the motor could be supported for the tests to be used, in this case it was created to place the motor suspended on a flat surface, in this way

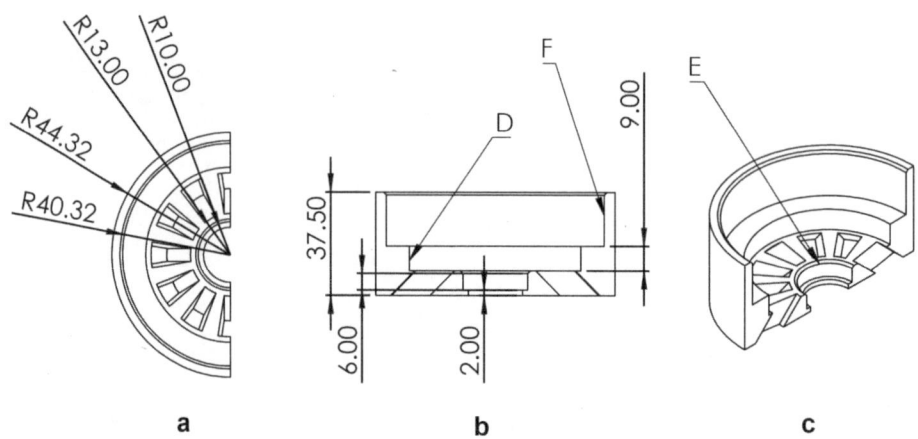

Fig. 7.24 Prototype one, housing

this base was designed with different operations, As can be seen, there is a rectangle where the motor core will be assembled, which will be the static part of the motor, this rectangle has the holes *A*, which will be aligned with the holes of the core cables, since the rectangle is hollow, the cables can be led through and then removed through the cut *B*, also the support has a protrusion *C* that acts as a stop to restrict the core (Fig. 7.25).

Fig. 7.25 Prototype one, motor's base

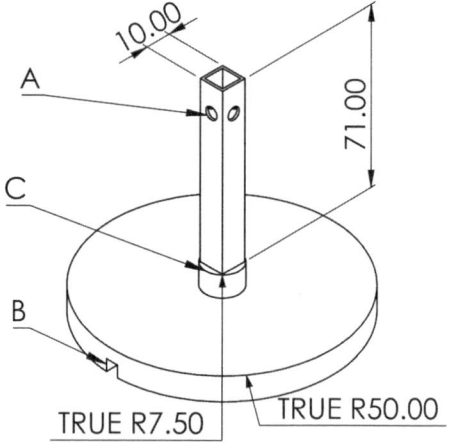

7.6   Prototype Two (18-20 halbach)

The second prototype was created for testing purposes for most of the dimensions to be used on a bicycle with the same number of coils and poles configuration as the first prototype, however, a halbach configuration was used to increase the motor torque (Fig. 7.26).

As can be seen, the second prototype has a different design than the first one, although they share the same configuration of poles and coils. This new prototype was designed on a 2-inch aluminum tubular shaft, which can be found in any store of this material, due to the fact that the shaft is not machined, bushings were used to center the housing and by default the magnets, the core of the coils is press fit into the tubular shaft as well as the bushings, In this way the core is secured to prevent it from slipping due to counter electromotive forces, in the same way as the previous design, the covers are used as a base for the core of the magnets giving support and structure to place it in place, in addition, it

Fig. 7.26 Prototype two, exploded view

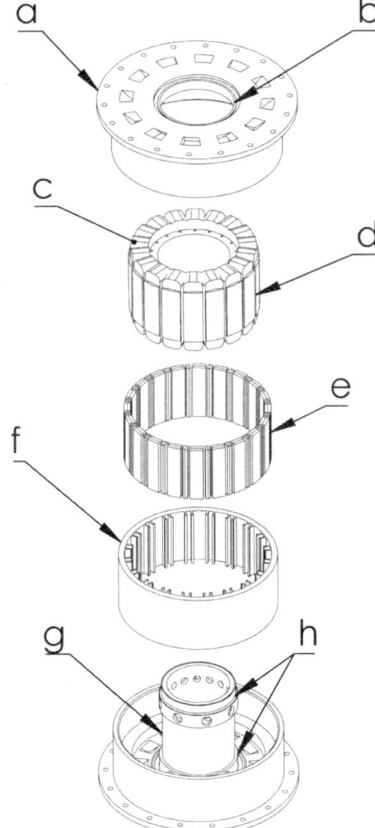

was designed to be mounted on a bicycle, having a support with slots to place the spokes of a bicycle.

7.6.1 Magnets Core

As in the previous design, for this design it is necessary to start with the materials that will not be manufactured, such as magnets, however, this design has a different configuration than the previous one it is called halbach, this was described before, the main advantage of this configuration compared to the normal one is its magnetic focus, by having a more focused field concentrates a greater magnetic flux in a specific point and therefore a greater separation between poles, this effect generates an arrangement that increases the angular rotation because it is more separated, if it is taken into consideration that this force will be added as the poles pass, it gives us a motor with a higher cruising speed. In addition, another type of bearings will be used, in this case they will have an internal diameter of 50 mm, external diameter of 62 mm and 7 mm in height.

In the formula seen previously the number of magnets to be used was placed, this is because the number of magnets is equal to the number of poles, however, in this case it is different, therefore, now it must be replaced in the formula:

$$a = \frac{360°}{n_p} = \frac{360°}{20} = 18°$$

where a is the opening angle of the sketch and n_p is the number of poles to be use.

Figure 7.27 shows the idea of the sketch of the magnets core, if we know the measures of the magnets, it is known that a cylindrical piece is required to contain the magnets, and the magnetic field in HALBACH arrangement, which consists of placing magnets on the sides of the main magnet to push the magnetic field of this and thus focus its magnetic field to the center, then 2 magnets in pairs of different measures are required to achieve this effect, the main magnet will have the measures of $4 \times 10 \times 40$ and the focus magnet will be $3 \times 5 \times 40$, it is important that the magnets are magnetized in the correct way explained before.

where a sketch is designed containing three circles, the cylinder contained in the outer and middle circles is extruded to obtain the rigid part of the core of the magnets, while the inner circle with respect to the middle will be used to generate the bases of the magnets, the sketch has different measures which place the sketch in the status of fully defined, remembering that it is necessary to define the Angle and the maximum distance of the sketch that will be applied to the circular matrix operation. The sketch in question will have the functionality to hold the main magnet, because the secondary magnet will be pushed backwards by interaction between magnetic fields. In addition, the centrifugal force added to the magnetic force will generate that the secondary magnet is strongly restricted in the correct position, in addition to this, the 3D printing was printed so that

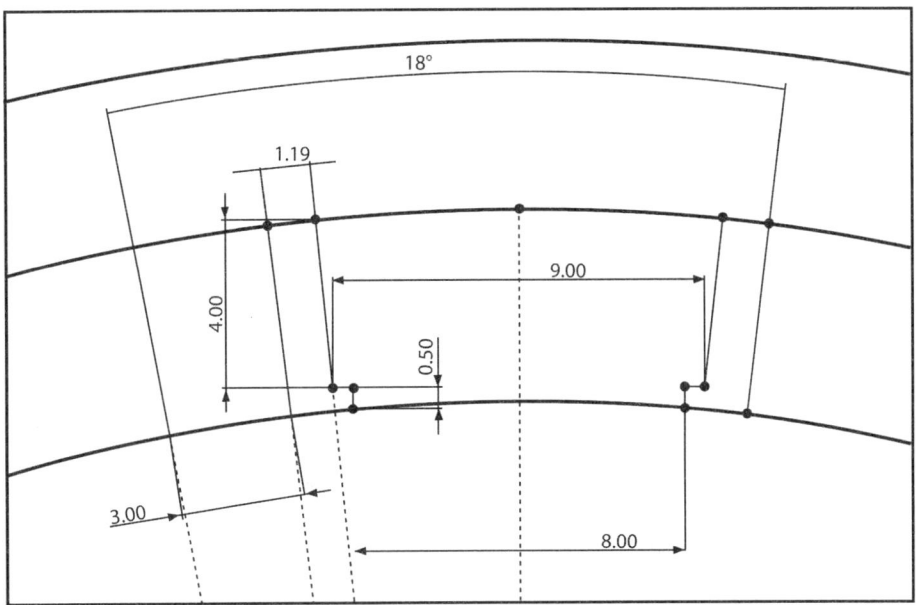

Fig. 7.27   Prototype two, sketch of magnet's base

the secondary magnet is trapped between the two bases of the main magnet, generating a greater restriction to the secondary magnet.

Remembering that this arrangement comes in pairs and the secondary magnet of the second arrangement will help the first one to focus the magnetic flux, as shown in Fig. 7.28.

Figure 7.29 (A) shows a top view of the part, as well as its different radial dimensions and its diametral exterior, and (B) shows an isometric view of the part where its height can be appreciated.

7.6.2   Coils Core

The core of the coils was designed on a 2-inch tubular shaft, therefore, this core design has two restrictions, it must be smaller than the minimum diameter of the magnets on the outside and larger than the tubular on the inside (Fig. 7.30).

This design has a hole in the center of the piece to put pressure on the 2-inch tube and an outer radius smaller than the minimum radius of the halbach magnet arrangement, due to the fact of being a halbach magnet arrangement the minimum diameter increases by placing a larger number of magnets compared to the normal arrangement, which is why, although the same poles were used, the minimum diameter is larger.

Fig. 7.28 Halbach magnet's
flux

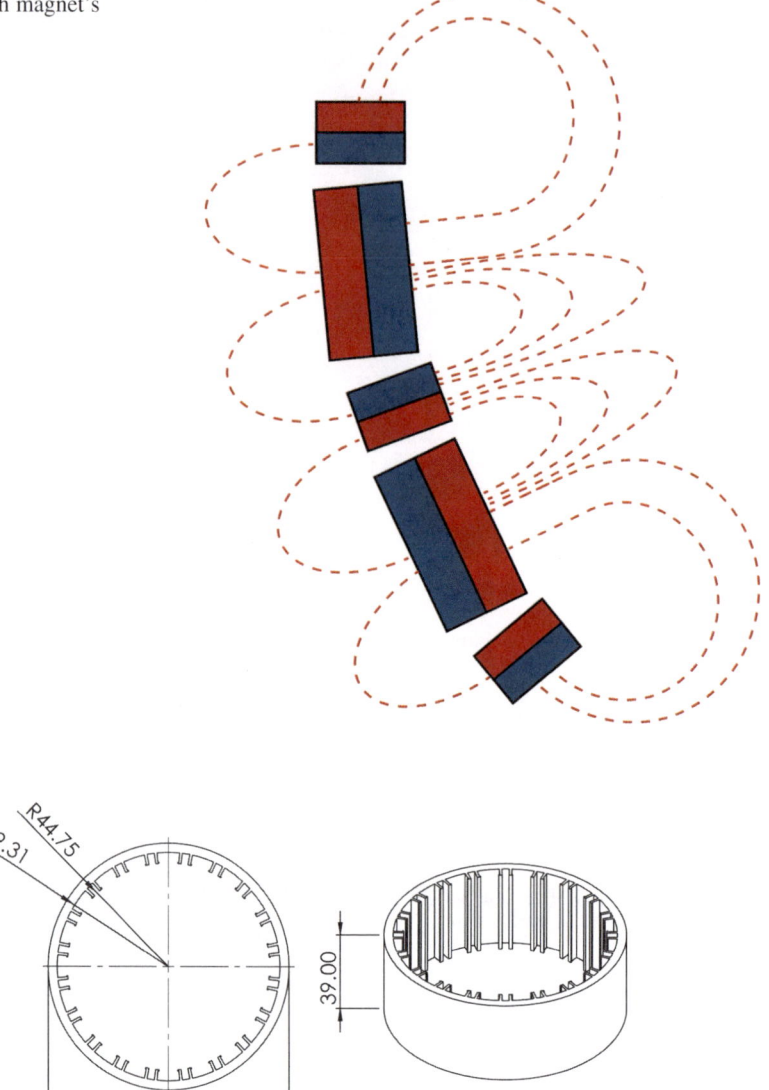

Fig. 7.29 Prototype two, magnets' base

7.6.3 Bushings

The bushings, although they may not seem of great importance are an essential part of the design, their main function is to keep certain parts separated, that is why they are simple

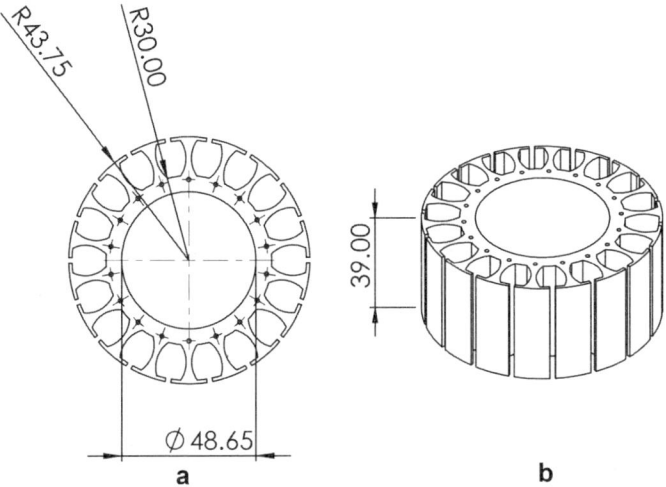

Fig. 7.30 Prototype two, core's coils

to design, they are usually a cylinder with a certain length, in addition, in this design it must be taken in consideration that the bearing does not fit perfectly in the tubular (Fig. 7.31).

For this design was proposed to use a separation of 10 mm since the housing do not touch the coils, in this way it will not interfere with the lid when rotating, also used a few holes that will go to the shaft (which will be drilled) to pass the cables inside the shaft and included a small ledge that will go inside the bearing in order to be well adjusted and this will depend on the clearance of the bearing with respect to the tubular, It is necessary a machining to achieve a correct adjustment of the bearing, although in this case it is solved by 3D printing, it should be noted that this design will not have the same strength as one of metal or a harder material, such as nylalloy.

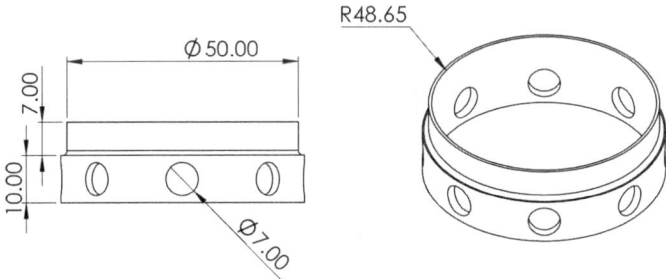

Fig. 7.31 Prototype two, bushing

7.6.4   Housing-Covering

In this design the cover will be the lid of the motor, as well as the motor housing. The main function of the cover in this design is to give structure to the magnets, that is to say, to accommodate the magnets in the right place, and for them it will be necessary to make a structure that will be supported by the bearings, since this will rotate (Fig. 7.32).

This part has several functionalities, the priority is to give structure to the base of the magnets, for this a hole c to the size of the bearing will be used and the part was designed with a ledge d that will restrict the movement of the core of the magnets, calculated to center them with respect to the core of the coils. In addition, square cutouts were added for ventilation. This sketch was made to be mounted on a bicycle rin, that is why the slots f where the spokes will be placed were added.

7.7   Motor Sheet I, 39 Teeth, 40 Poles

The unbalance of the motor is indispensable to achieve the rotational movement of the motor, speaking in terms of the arrangement of coils and magnets. However, the more unbalanced it is, the less power it will be able to convert from electrical to mechanical. The way to determine how balanced the motor is by means of the ratio between poles and coils, the further away from 1 the more unbalanced the motor will be, the only way

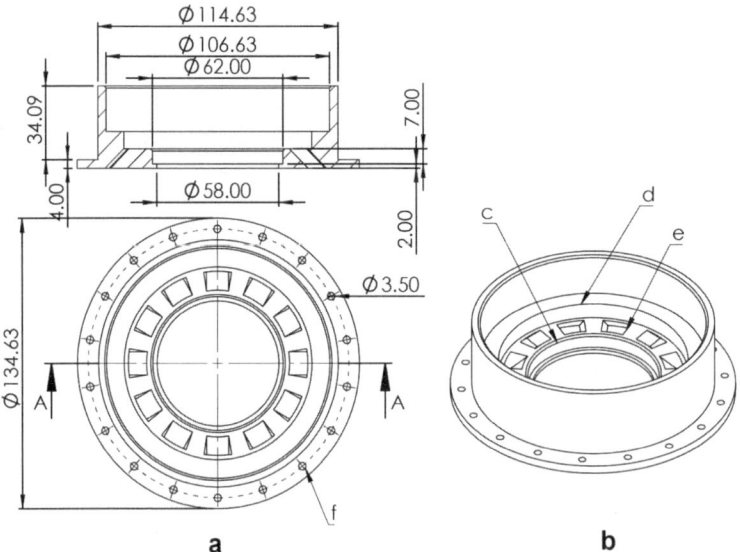

Fig. 7.32 Prototype two, housing-lid

to change that is by modifying the number of magnets and coils. At this point the 39-40 motor is an arrangement that is capable of being built.

Figure 7.33 shows an exploded view of all the parts that make up the motor of the third prototype, this prototype was designed to use machining techniques other than printing, using a metal shaft turned and drilled to size, that process can be ordered in some machining workshop.

This prototype was designed to observe the differences between a halbach magnet arrangement and a normal one, so a design had to be made that could have both configurations.

Fig. 7.33 Motor sheet I, exploded view

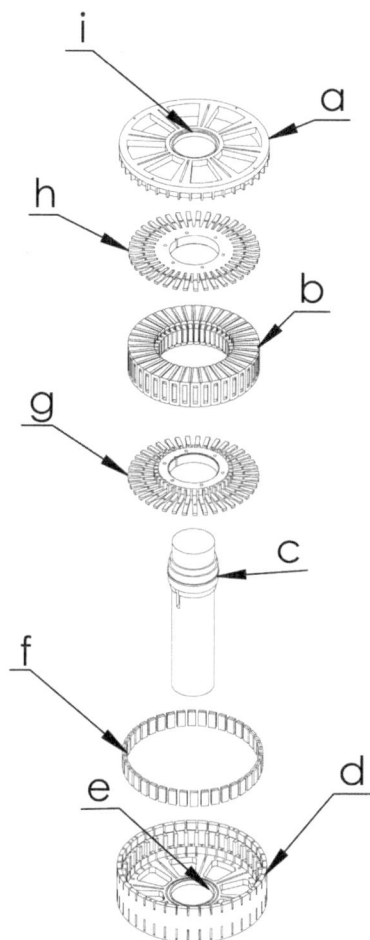

7.7.1 Restrictions

For this design, the 2 previous restrictions were used (the size of the bearings and the size of the magnets), in addition, the measurements of the electrical steel sheets were taken into consideration, since they need to have the minimum diameter separation, respect the space of the shaft and also have enough separation between them to give the necessary turns in the coils, for this reason, the 39-40 configuration is better than the 18-20, since it has a much larger minimum diameter of the magnet arrangement.

Figure 7.34 shows the dimensions of the laminate that will be used for the core of this motor, the core design must restrict all the degrees of freedom of this plate in order to keep it in place, the easiest way to do this is to use the way it is designed to be placed, through the hole that has to pass a bolt and make a small groove on one side to prevent it from rotating, once stacked all movements will be restricted.

It is worth mentioning that the laminations must be isolated from each other to avoid parasitic currents, which generate magnetic fields contrary to the electromotive force.

In Fig. 7.35 you can see the space E where the coil will be roll, this space should be sufficient to accommodate the threads corresponding to the calculation plus an extra space that will be used to easily wind the threads. This space should generally be twice the size of the thickness of the thread so that it can easily pass to the beginning of the tooth.

Fig. 7.34 Sheet I

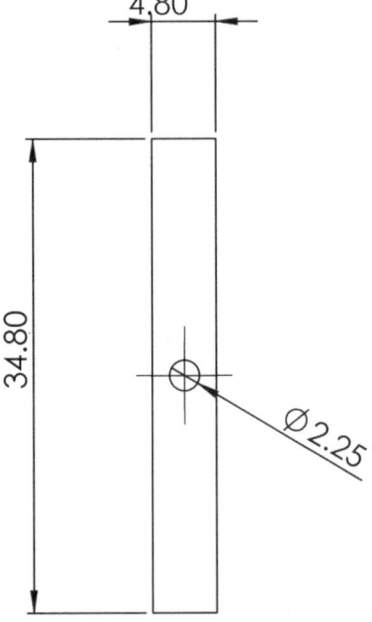

Fig. 7.35 Space between coils

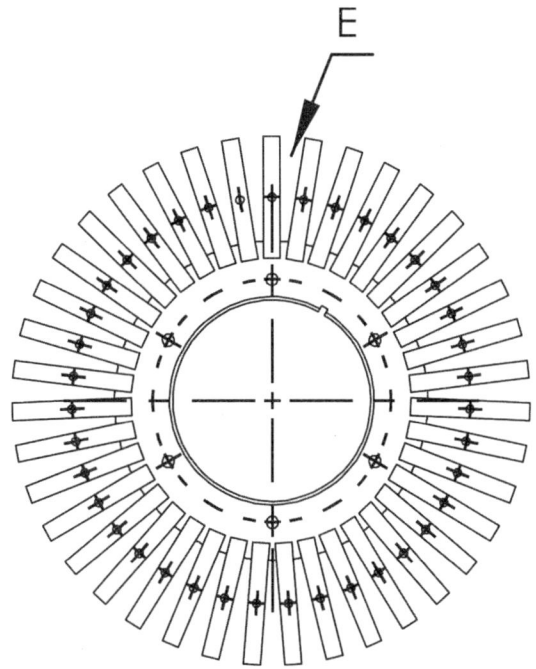

7.7.2   Magnets Housing

The core of the magnets was the first to be designed in previous versions, because it contains the only parts which will not be printed, however, as there are more parts which are prefabricated, it is necessary to make an adjustment. The first thing to do is to identify the minimum range that the core of the coils can have, once this range is identified, the number of poles can be calculated using the formulas described above or alternatively, iterate with various coil-poles configurations to obtain a minimum diameter that is within this range. Another option would be to separate the magnets, putting a configuration in a larger diameter than the minimum, however, this decreases the efficiency of the motor exponentially (Fig. 7.36).

The sketch to be able to define the minimum diameter is the same as the second prototype, except for the angle, this angle will have to be adjusted to the pole configuration that this motor will have, in this case it will be a 40 pole configuration, remembering that in the halbach configuration the number of poles must be used instead of the number of magnets, if it is substituted in the formula the following is left:

$$a = \frac{360°}{n_p} = \frac{360°}{40} = 9°$$

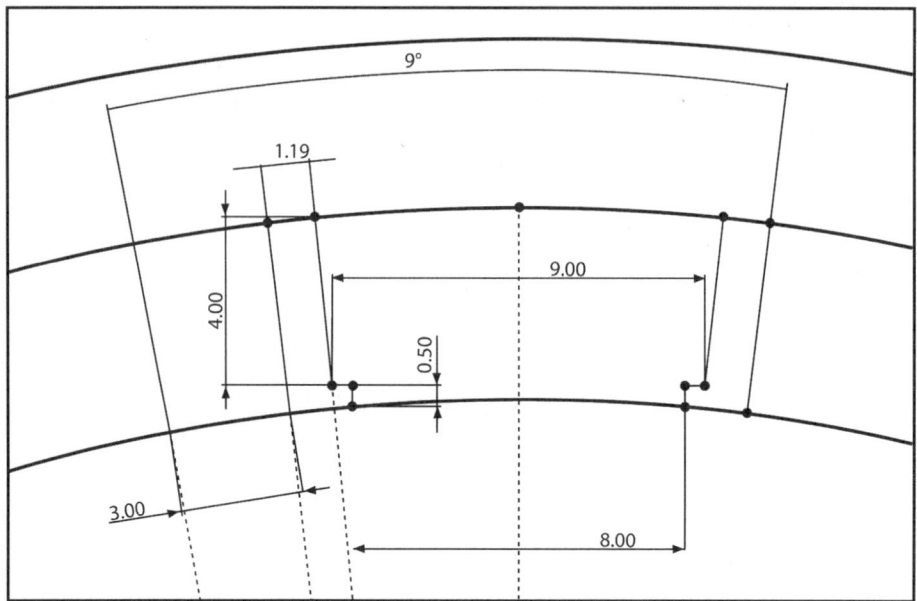

Fig. 7.36 Motor sheet I, minimal diameter of magnets

The way of manufacturing in a 3D printer gives a wide range of design options when it comes to shapes. To take advantage of this process, a solid part containing the magnet core and the housing was designed. Figure 7.37 shows the top view of this model, where you can see a detailed view of the slot for the magnet, this slot is designed so that the magnet enters under pressure, being difficult to remove from it, for this, the surface where the magnet touches has a hole past *c* where you can insert a rod to push the magnet. In addition, in the upper part of this piece there are slots *d*, which are aligned with the ledges of the lid, so that when joining the pieces, a strong enough union is obtained to prevent them from separating, in addition to this, there are holes *g* which are aligned with holes in the lid to screw the lid with the housing. Finally, there is the structure *b* in which the bearing is placed.

7.7.3 Core of the Coil

The prototype 3 was thought to be built with electrical steel sheets, these are sheets used for transformers and have interesting characteristics compared to other types of materials, for a material to be good to be used in a coil tooth, 3 things are needed mainly. The first is that the material has a low electrical resistance in the direction of the vector resulting from the dot product between the motion vector and the magnetic flux vector. The second

Fig. 7.37 Motor sheet I, magnet's housing

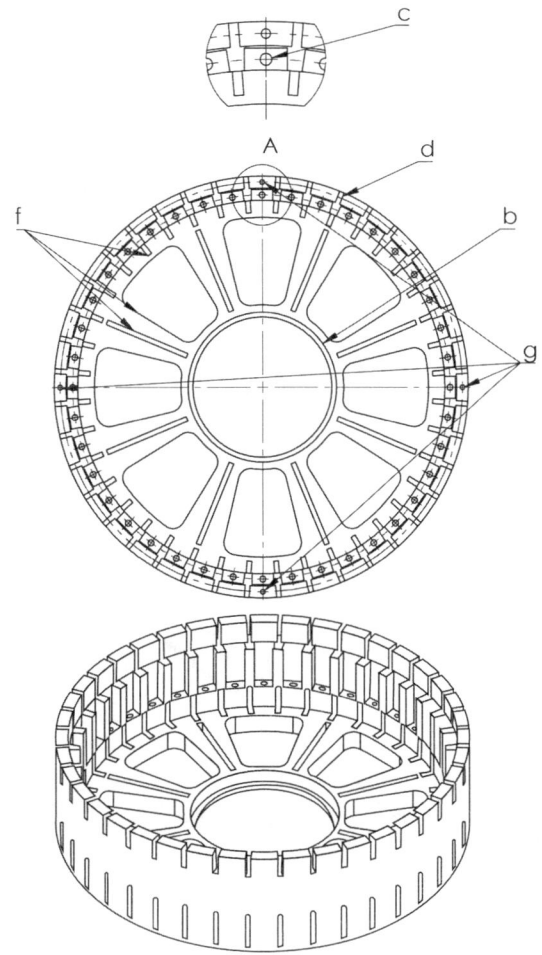

is that it has a low magnetic reluctance (magnetic resistance) and that it demagnetizes quickly when the magnetic field is no longer applied.

Electrical steel has very good magnetic conduction and demagnetizes quickly, however, it is also a very good conductor of electricity, although this can be reduced by laminating and insulating the material in the direction of the vector resulting from the dot product between the motion vector and the magnetic flux vector, shown in Fig. 7.38 as the I vector.

For axial motors the vector I will be in the direction of the motor axis, either up or down depending on the motor rotation. Therefore, laminating the material in this direction will drastically increase the efficiency of the motor, in the case of an electrically conductive material such as electrical steel.

Electrical steel laminations are used for transformers, because transformers must have similar characteristics to motors and their wide variety of applications to most industrial

Fig. 7.38 Motor sheet I, coil

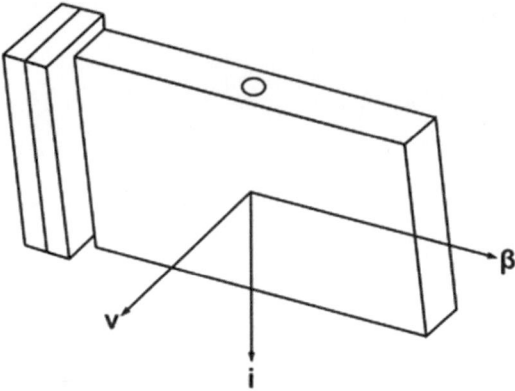

and domestic machines, this type of laminations are very commercial, so they are a good candidate for prototyping motors (Fig. 7.39).

There are two types of sheets to create the motors, commercially speaking, the type *I* and type *E*. The measures of the type *I* sheet are specified in the Fig. 7.34 indicated as

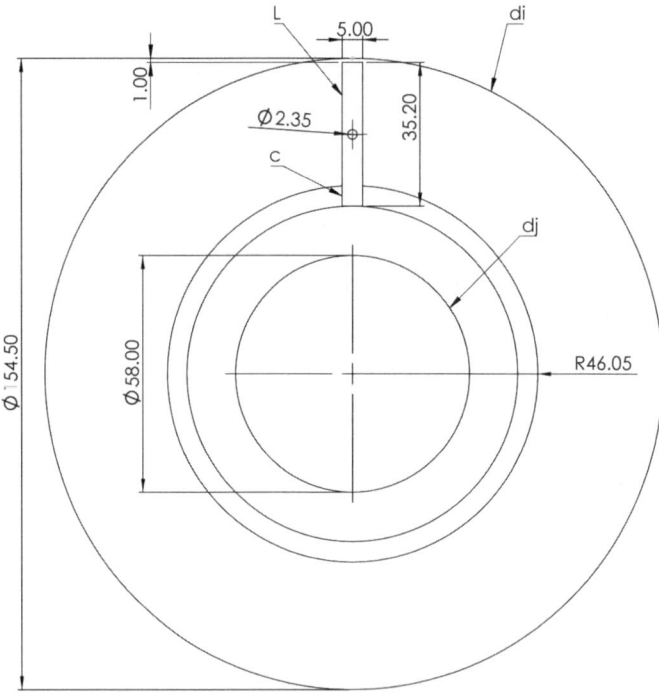

Fig. 7.39 Motor sheet I, sheets I

the picture *L*, in addition they have a thickness of 0.65 mm and they have a hole through which they can be put in place. The assembly of the transformers consists of overlapping the set of these two pieces, in such a way that piece *E* is on top of piece *I* centering with them holes, the method to assemble the motor will be similar, using this hole to put the pieces in place, however, this will be done using piece *I* for this prototype.

Figure 7.39 shows the measurements that were used to design the motor core, in addition to highlighting some important parts such as the joint highlighted with the letter *c*, it can be seen how it is planned so that the sheets rests on the core and is completely restrained by the through bolt mentioned above. In addition, it shows the *di* which is the minimum diameter of the magnets and the first dimension to be placed, the second dimension placed is the *dj*, this being the diameter of the shaft, from this sketch you can extrude the different base ledges that will be in the core.

Small brushless motors usually use I-shaped teeth, however, this causes a coggin effect which in turn causes unwanted vibrations. For larger motors it is better to use teeth that cover as much surface area as possible, this slows down the coggin effect considerably, resulting in a smoother running motor.

Figure 7.40 shows the design of the finished core, this was designed in two parts, one will be placed on one side of the shaft and the other on the opposite side, the only difference between the two will be the cut of the wedge *c* which will be on the opposite side of the front plane. The piece has a cut in the shaft hole *a*, which is used to place it in the shaft, since the shaft has this ledge that will serve as a core lock to prevent it from moving axially. In addition, it has the hole *c* where a wedge will be placed to prevent radial movement. On the other hand, the piece has the holes *b* that will serve for the through bolts, these bolts will pass through the two parts of the core, joining both pieces. The ledge *e* is the part where the electrical steel sheet will be supported and was designed along the entire length of the sheet in order to protect the copper wire from scratching with this same sheet, also the ledge *e* has a cut *f* used at the time of winding the cable so that it does not slip towards the rotor and cause a failure.

Fig. 7.40 Motor sheet I, coils' core

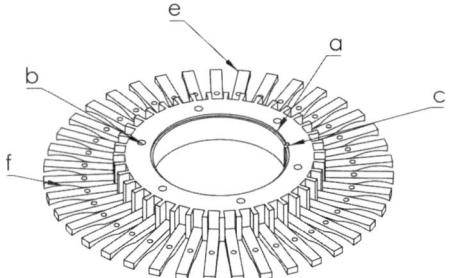

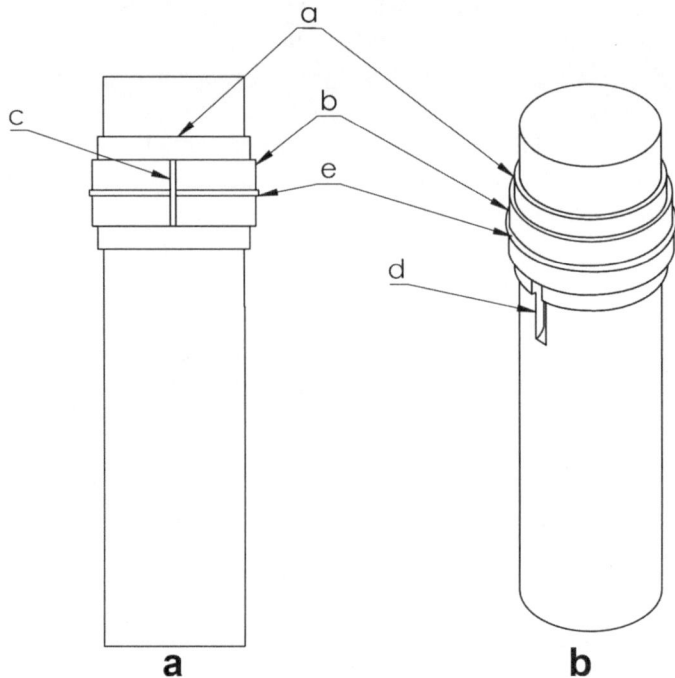

Fig. 7.41 Motor sheet I, shaft

7.7.4   Axis

The shaft was designed to be machined by turning, it has different levels and cuts where the bearings and the core of the coils will be supported. Level a will be used to seat the bearing and restrain it axially, level **b** will be used to place the cores in place generating restrictions with level **e** that will act together with slot **a** of the core shown in Fig. 7.41, and also with slot **c** that will act together with slot **c** of the core shown in Fig. 7.41. Finally, the shaft has **a** slot **d** used to obtain the cables from the inside of the motor.

7.7.5   Lid

This part of the motor was designed to be joined to the press-fitted housing part, retaining the bearing, with the proper size to keep everything in place (Fig. 7.42).

The lid of this prototype, has a peculiar design, the protuberances **d** were designed to be able to press into the slots **d** of the housing, it also has the holes **b** which will be aligned with the holes **g** of the housing to be screwed, since 3D printing is soft compared

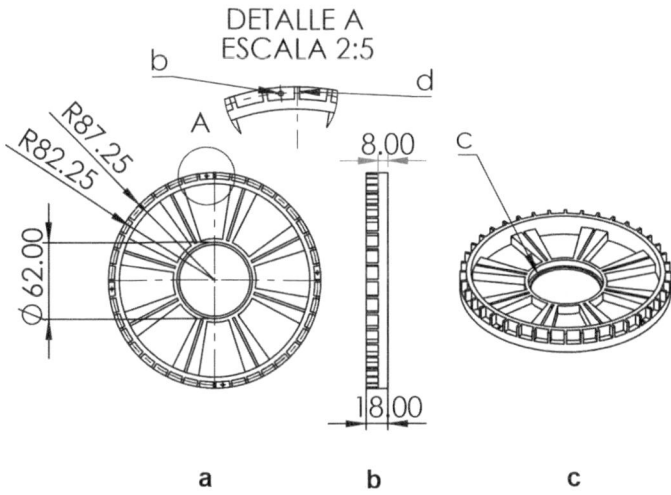

Fig. 7.42 Motor sheet I, lid

to metal a pin can easily make its way over it, being screwed firmly. Finally, you have the groove **c** in which the bearing will be mounted.

7.8   Motor Sheet E

The shape of the tooth, as well as the material with which it is made, greatly influences how the motor performs. One of the effects to take into consideration when designing the motor is the coggin effect, this effect is caused due to the magnetic attraction of the magnet with the electric steel core, which causes the motor to jam at those points where the magnet is attracted with more force, causing the motor to generate vibration, this can be reduced by increasing the surface of the tooth closest to the magnet, causing the attraction to disperse over the entire surface of the core (Fig. 7.43).

As in the previous prototype, the electrical steel profiles manufactured for the transformers were used for this prototype. These profiles will be the counterpart of the I-profile used in the previous prototype, which are called E-sheets. Like the previous profile, they can be purchased at a store that sells this type of products. These profiles will be cut at a distance of 16 millimeters in order to leave them to size for the design.

7.8.1   Embedded Core

3D printing is an interesting way to manufacture objects, the generation of the pieces depends on the arrangement of the piece, since it is possible to print a piece in different

Fig. 7.43 Sheet E

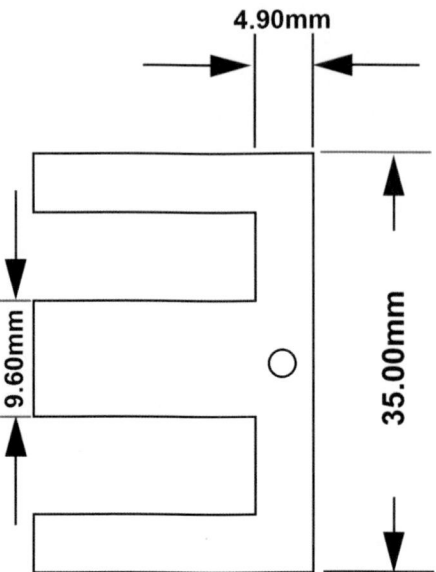

planes that exist, the general idea when arranging a piece is to put it so that the lower layers support the upper layers.

Figure 7.44 shows two ways to place the piece in item *A* shows the part with less surface of the piece pointing down, this generates that some parts are floating in the air which causes these parts to fall to the printer bed, in item *B* is observed as all layers have a support layer before where the hot filament will be placed and fused, thus generating a solid and well printed piece. The designs for 3D printing can usually be adapted to print them, however, sometimes there is no way to do it with a solid piece so there are four options. The first and easiest is to use the printing software to put supports, the second is to use the design to adapt the part, the third is to print two separate parts and join them together using heat, screws, glue, etc... The fourth is to use a printing technique called embedded print, this consists in printing a part and stop it in a specific layer to insert another part with the purpose of printing the following layers over the inserted part.

Figure 7.45 shows the core, which is made with two of the four techniques mentioned above. One of them is the printing of the cores *b* and *d* separately, the core *b* is a thin sheet with ledges, which enter the holes of the core *d* under pressure.

In addition, the bushing that will be used to place it on the metal shaft will be embedded the piece of the two cores already assembled, at the time of printing this the printer is configured with the purpose of stopping just before starting the upper cylinder, then the core is embedded and continues with the printing, the following layers will be printed on the core piece. At the end of the printing the parts *b* and *d* will be embedded inside the hub *a* as one solid part.

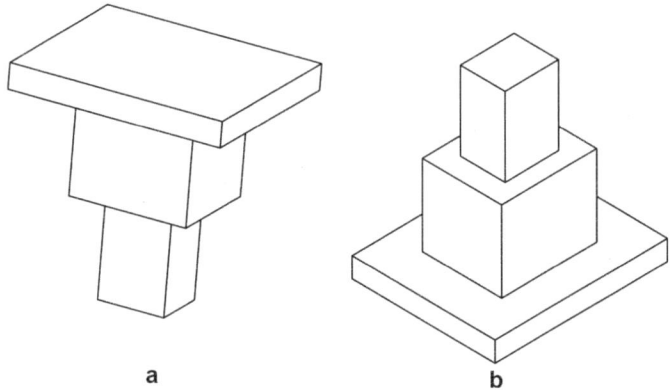

Fig. 7.44 3D support

Fig. 7.45 Motor sheet E,
coil's core

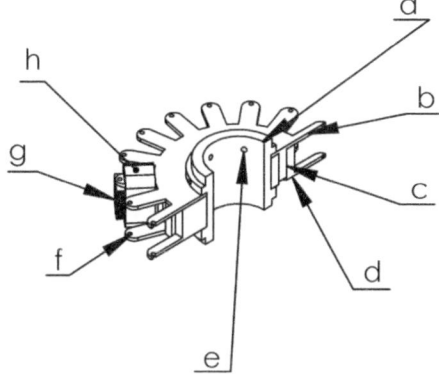

Both parts have supports for the sheets, as well as a bore *f* from which the sheets will be riveted once in place to form the tooth core *g*, where the coil *h* is going to roll.

7.8.2   Assemble

Figure 7.46 shows the motor assembly cut in half and each part moved individually by certain degrees, this was done with the purpose of knowing the internal assembly of the motor and to see how it is distributed internally.

In the assembly of the prototype with *E* sheets, it can be observed that the embedded core *b* is the one that holds all the parts, it has a ledge *h* to separate the bearings, it has bases with a hole through to rivet the *E* sheets and thus form the teeth *e* where the coil *c* will be wound, besides holding them on the other side, in addition, the embedded bushing is designed to enter under pressure in the shaft *d*.

Fig. 7.46 Motor sheet E,
assemble

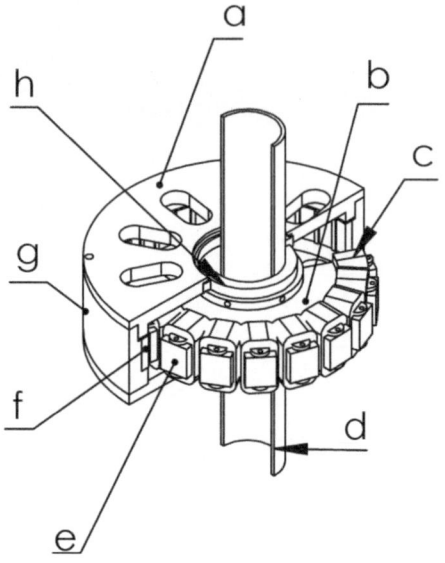

The housing *g* is designed to be screwed with worm screws together with the top and bottom cover *a*, as well as having the base for the magnets in the same piece.

7.8.3 Motor Sheet E Lid

The cover has a design which fits into the housing, with the ledge *d* mentioned in the image as a centering device, also has as the covers of the other designs a hole *c* where the bearing is housed, also to pass the shaft *I* place in the design the hole *a* where you can put the shaft, on the other hand, the *b* holes passed that are in the previous circumference are designed to place a worm gear that crosses the motor from side to side (Fig. 7.47).

Fig. 7.47 Motor sheet E, lid

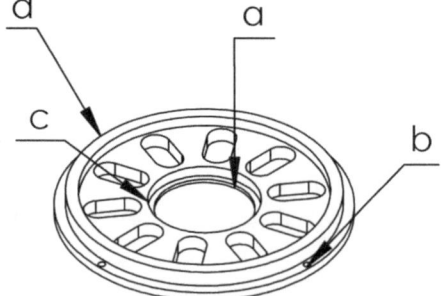

Fig. 7.48 Motor sheet E, housing

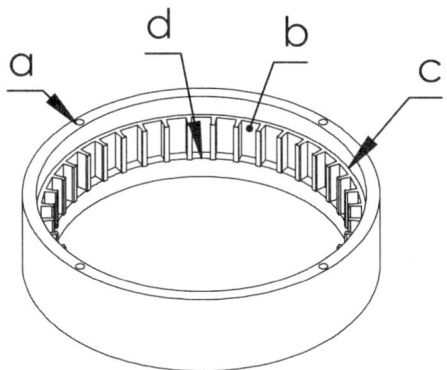

7.8.4 Magnet's Base

Figure 7.48 shows an isometric view of the housing, where the main features that make it functional can be observed, such as the holes passed through, where the worm screws will be placed to hold the bottom cover and the top cover. The design also has the bases for the magnets, which in this case will be 20 × 10 × 5 mm, in a configuration of two magnets per pole, i.e., in the arrangement two magnets with north polarity will be placed in the center followed by two south polarity. The reason for this arrangement is due to the width of the coil tooth since the *E* sheet is 16 mm wide as it was re-cut from its total 35 mm. The ledge *c* is located just where the magnets end and serves to make the ledge.

References

1. TheGoofy www.laimer.ch, "600 Watt, 3d-printed, Halbach Array, Brushless DC Electric Motor," *Instructables.* https://www.instructables.com/600-Watt-3d-printed-Halbach-Array-Bru shless-DC-Ele/
2. H. Raich, P. Blümler, Design and construction of a dipolar Halbach array with a homogeneous field from identical bar magnets: NMR Mandhalas. Concepts. Magn. Reson. Part B Magn Reson Eng **23B**(1), 16–25 (Oct.2004). https://doi.org/10.1002/cmr.b.20018
3. SKF Gruop, "Rolling bearings," Oct. 2008. Available: https://cdn.skfmediahub.skf.com/api/pub lic/0901d196802809de/pdf_preview_medium/0901d196802809de_pdf_preview_medium.pdf